TELEPEN

9023072 8

Q'

D1513304

Contributions to Nephrology

Vol. 12

Series Editors
G.M. Berlyne, Brooklyn, N.Y.; *S. Giovannetti,* Pisa, and
S. Thomas, Manchester

Editorial Board
J. Brod, Hannover; *J. Churg,* New York, N.Y.;
K.D.G. Edwards, New York, N.Y.; *C.J. Hodson,* New Haven, Conn.;
S.G. Massry, Los Angeles, Calif.; *R.E. Rieselbach,* Madison, Wisc., and
J. Traeger, Lyon

S. Karger · Basel · München · Paris · London · New York · Sydney

Vasoactive Renal Hormones

Volume Editors
G.M. Eisenbach and *J. Brod,* Hannover

53 figures and 14 tables, 1978

S. Karger · Basel · München · Paris · London · New York · Sydney

Contributions to Nephrology

This volume contains selected papers presented at the 5th Nephrological Symposium, Hannover, June 10–11, 1977

Vol. 9: Recent Advances in Renal Research. Contributions from Japan.
Kobayashi, K.; Maeda, K., Nagoya, and Oshima, K., Tokyo (eds.)
VI + 130 p., 54 fig., 43 tab., 1978. ISBN 3–8055–2826–4
Vol. 10: Toxic Nephropathies. Migone, L. (ed.)
VIII + 114 p., 49 fig., 23 tab., 1978. ISBN 3–8055–2832–9
Vol. 11: Unilateral Renal Function Studies.
Bianchi, C., Pisa, and Blaufox, M.D., New York, N.Y. (eds.)
X + 210 p., 77 fig., 2 cpl., 23 tab., 1978. ISBN 3–8055–2858–2

Cataloging in Publication
Nephrological symposium, 5th, Hannover, 1977. Vasoactive renal hormones /
volume editors, G.M. Eisenbach and J. Brod. – Basel, New York : Karger 1978
(Contributions to nephrology ; v. 12)
Contains selected papers presented at the symposium held in Hannover,
June 10–11, 1977
1. Angiotensins – congresses   2. Kallikrein – congresses   3. Renin – congresses
4. Prostaglandins – congresses
I. Brod, Jan, ed.   II. Eisenbach, Georg M., ed.   III. Title   IV. Series
Wl CO778UN v. 12   WK180 N439 1977V
ISBN 3–8055–2839–6

All rights reserved.
No part of this publication may be translated into other languages, reproduced or utilized in any form or by any means, electronic or mechanical, including photocopying, recording, microcopying, or by any information storage and retrieval system, without permission in writing from the publisher.

© Copyright 1978 by S. Karger AG, 4011 Basel (Switzerland), Arnold-Böcklin-Strasse 25
Printed in Switzerland by Thür AG Offsetdruck, Pratteln
ISBN 3–8055–2839–6

NIH88
QUBML
- 9 SEP 1982
BELFAST

Contents

Contents

Contr. Nephrol., vol. 12, pp. 1–4 (Karger, Basel 1978)

Introduction

Jan Brod, M.D., Dr. Sc., FRCP

Hannover Medical School, Hannover

The idea that, in addition to its excretory function, the kidney is also an incretory organ, was suggested for the first time almost a century ago by *Brown-Séquard and d'Arsonval* (1892). They thought that uremia is a consequence of the failure of this internal secretion because it may be absent in cases of anuria. That this argumentation is not valid has been, of course, recognized long ago and had, at the times I studied medicine just over 40 years ago, anybody the audacity to answer the question about the function of the kidney by the statement that it is endocrine, he probably would have been failed without any further discussion. This in spite of the fact that 1898 *Tigerstedt and Bergmann* extracted a substance from the rabbit renal cortex raising the blood pressure of another rabbit upon intravenous injection. Extracts from other organs had no such activity. They called the pressor principle 'renin'. This fact, though confirmed 11 years later by *Bingel and Strauss* (1909), was completely forgotten in the three decades to follow and it is astonishing to note that it was never considered by *Volhard* and his school during the quarter of a century of their futile search for a vasoconstrictor substance in the blood of patients with 'pale' hypertension.

It was only after *Goldblatt et al.* (1934) had published the results of their famous experiments and after the demonstration that the clipped kidney can still produce hypertension when its nervous connections with the body were completely severed, that renin was revived. Almost without delay *Page* (1939) and his colleagues in Indianapolis and *Braun-Menéndez et al.* (1940) in Buenos Aires simultaneously and independently discovered that it is not renin but a product of its interaction with the α_2-globulins of the plasma that is the pressor substance proper. I do not intend to recapitulate here the history of the confusion that followed due to the dual nomenclature for the same thing these two groups have introduced, until in the late fifties the term angiotensin was agreed upon and until its octapeptide structure was recognized for the first time by *Peart* (1956). There is no doubt that angiotensin, even if not secreted as such by the renal cells, is formed by the action of renin − a true incretory product of the kidney − on the plasma globulins within the kidney and can be found both in the blood and in the lymph leaving the kidney. It is, thus, a true hormone. Few other substances attracted so much attention and incited so much excellent research in the past two decades. Per unit weight, angiotensin is the strongest blood pressure elevating agent known. This together with the fact that the kidney was linked with high blood pressure even since *Bright*'s days (1836) led to the hypothesis that an increased release of angiotensin, producing vasoconstriction, is at the basis of human hypertension. So far this simple belief was not substantiated. (1) There is no obvious reason why a morphologically healthy kidney at the initial phase of essential hypertension should release renin and form angiotensin in increased amounts; (2) a hyperangiotensinemia has never been encountered in uncomplicated human hypertension either of the essential or renal variety; (3) the hemodynamic pattern produced by angiotensin differs from that of both these varieties of hypertension; (4) antagonists do not affect the high blood pressure in benign hypertensives. Nevertheless, angiotensin has several important functions in the body: as shown in the pioneering work of *Genest et al.* (1960), *Laragh et al.* (1960) and *Gross* (1960), it stimulates the release of aldosterone and is, therefore, connected with the sodium metabolism and conversely the state of the sodium balance may affect the responsiveness of the vessels to angiotensin. It further affects the hypothalamic sympathetic centers (*Bickerton and Buckley,* 1961) and the release of catecholamines from the adrenal medulla (*Feldberg and Lewis,* 1964).

Yet another renal hormone or hormonal system has been detected in connection with hypertension. Since 1943, *Grollman* has been insisting that the removal of both kidneys can raise the blood pressure of the surviving experimental animal. After many years of systematic studies *Grollman*'s past associate *Muirhead* (1976) and his colleagues succeeded in convincingly demonstrating the presence of a 'blood pressure lowering hormone' in the renal medulla, probably stored and secreted by the interstitial cells. The lipid character of this substance suggests that it may be related or identical with prostaglandins found in the renal medulla. Of these, prostaglandin E_2 is strongly vasodilator whereas prostaglandin $F_{2\alpha}$ is recently being considered in connection with tubular reabsorption of sodium. The increased release of prostaglandin E into the renal blood in response to an infusion of catecholamines or angiotensin and the fact that it probably does not pass the lung barrier has led to its being considered as a local modulating factor. We are happy to have with us Prof. *McGiff* and Prof. *Lee,* who have carried out most of the pioneering work, to lead us through the maze of conflicting data, wild hypotheses and solid facts to a proper perspective.

These agents can at present hardly be discussed without touching upon yet another intrarenal biological system whose action is linked with that of the prostaglandins – the renal kallikrein-kinin system, whose importance has so vividly been stressed at the last International Congress of Nephrology in Florence by Prof. *Mills,* and whose leading protagonists Professors *Pisano* and *Nasjletti* are all here with us. The prostaglandins seem to be involved in various renal physiological processes, such as the regulation of renal medullary blood flow and in the action of ADH as will be recounted by Prof. *Douša.*

An international symposium of this size involves, of course, high costs which no medical school could bear without a most generous support from outside. This was obtained from the following houses: BASF, Ludwigshafen; Beiersdorf, Hamburg; Boehringer, Ingelheim; Boehringer, Mannheim; Ciba-Geigy, Frankfurt/Main; Deutsche Wellcome, Grossburgwedel; Fresenius, Bad Homburg; Giulini-Pharma, Hannover; Grünenthal, Stolberg; Gry-Pharma, Kirchzarten; Hoechst, Frankfurt/Main; Hoffman-La Roche, Grenzach; ICI, Heidelberg; Labaz, Düsseldorf; Lepetit, München; Madaus, Köln; Merck, Darmstadt; Nordmark, Uetersen; Paul-Martini-Stiftung, Frankfurt/Main; Pfizer, Karlsruhe; Rhöm-Pharma, Darmstadt; Sandoz, Nürnberg; Schering, Berlin. I would like to express to them our most sincere appreciation and gratitude.

References

Bickerton, B.K. and Buckley, J.P.: Evidence for a central mechanism in angiotensin induced hypertension. Proc. Soc. exp. Biol. Med. *106:* 834 (1961).

Bingel, A. und Strauss, E.: Über die blutdrucksteigernde Substanz der Niere. Dt. Arch. klin. Med. *96:* 476 (1909).

Braun-Menéndez, E.; Fasciolo, J.C.; Leloir, L.F., and Muñoz, J.M.: The substance causing renal hypertension J. Physiol., Lond. *98:* 283 (1940).

Bright, R.: Cases and observations illustrative of renal disease accompanied with the secretion of albuminous urine. Guy's Hosp. Rep. *1:* 338 (1836).

Brown-Séquard et d'Arsonval: Des injections sous-cutanées ou intra-veineuses d'extraits liquides de nombre d'organes comme méthode thérapeutique. C.r. hebd. Séanc. Acad. Sci., Paris *114:* 1399 (1892).

Feldberg, W. and Lewis, G.P.: The action of peptides on the adrenal medulla release of adrenaline by bradykinine and angiotensin. J. Physiol., Lond. *171:* 98 (1964).

Genest, J.; Nowaczynski, W.; Koiw, E.; Sandor, T., and Biron, P.: Adrenocortical function in essential hypertension. In *Bock and Cottier* Essential hypertension. An International Symposium, p. 126 (Springer, Heidelberg 1960).

Goldblatt, H.; Lynch, J.; Hanzal, R.F., and Summerville, W.W.: Studies on experimental hypertension. I. The production of persistent elevation of systolic blood pressure by means of renal ischemia. J. exp. Med. *59:* 347 (1934).

Grollmann, A. and Rule, C.: Experimentally induced hypertension in parabiotic rats. Am. J. Physiol. *138:* 587 (1943).

Gross, F.: Adrenocortical function and renal pressor mechanism in experimental hypertension; in Essential hypertension. An International Symposium, p. 92 (Springer, Berlin 1960).

Laragh, J.M.; Angers, M.; Kelly, W.G., and Lieberman, S.: Hypotensive agents and pressor substances: The effect of epinephrine, norepinephrine, angiotensin II and others on the secretory rate of aldosterone in man. J. Am. med. Ass. *174:* 234 (1960).

Muirhead, E.E.; Rightsel, W.A.; Lerch, B.E.; Byers, L.W.; Pitcock, S.A., and Brooks, B.: Anti-hypertensive lipid from tissue culture of renomedullary interstitial cells of the rat. Clin. Sci. molec. Med. *51:* 287 (1976).

Page, I.H.: On the nature of the pressor action of renin. J. exp. Med. *70:* 521 (1939).

Peart, S.W.: The isolation of hypertensin. Biochem. J. *62:* 520 (1976).

Tigerstedt, R. und Bergmann, P.G.: Niere und Kreislauf. Skand. Arch. Physiol. *8:* 223 (1898).

Prof. Dr. *J. Brod,* DrSc, FRCP, Medizinische Hochschule Hannover, Departement für Innere Medizin, Abteilung für Nephrologie, Karl-Wiechert-Allee 9, Postfach 61 01 80, *D–3000 Hannover 61* (FRG)

Renin — Angiotensin

Contr. Nephrol., vol. 12, pp. 5–15 (Karger, Basel 1978)

Intra-Renal Factors in Renin Release

W.S. Peart

Medical Unit, St. Mary's Hospital, London

The last 10 years has seen a very large amount of work devoted to defining the factors concerned in renin release. Using a variety of physiological, pharmacological, possibly toxicological, and certainly pathological approaches, it has been possible to define most of the stimuli that have an effect within the kidney, and I will attempt to give my interpretation of the position.

Major consideration must still be given to the afferent arteriole since the juxtaglomerular cells containing renin lie within its walls just before the glomerulus (23, 24, 41), and the morphology of this area has been well studied (21) and one of the important points is the rich innervation of this region by predominantly sympathetic nerve terminals which can be demonstrated using fluorescence (18, 34) as well as by electron microscopy (4). The former is obviously more specific in terms of noradrenaline content and identification. Since *Tobian* (50) proposed the juxtaglomerular arteriole stretch or pressure hypothesis of renin release, there have been many experiments which test out the hypothesis and very few which refute the initial concept. The initial major problem was that any experiment which in order to change stretch or pressure involved a change of flow, would also influence the glomerular filtration rate and the composition of the urine, which, allied to the morphological consideration of the juxtaposition of the macula densa in the distal convoluted tubule with the afferent arteriole (24), led to the macula densa hypothesis which has been so strongly supported by *Thurau et al.* (48) and *Thurau and Schnermann* (49). This has led directly to efforts to dissociate the two possible components, as in the experiments of *Davis* (13) with the non-filtering kidney, presumably leaving the vascular effects in isolation, and the very direct approach to changes in the composition of the fluid arriving at the macula densa, with an emphasis on chloride flux (43, 48). Most of the experiments which relate to the vascular component, at least in the mind of those who carry out the experiment, involve at least in theory changes in urinary composition which could affect the macula

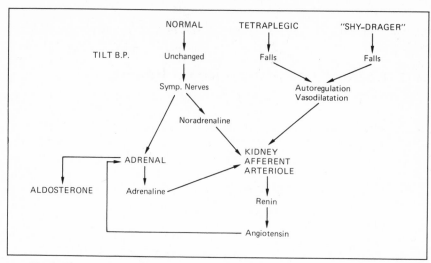

Fig. 1. The suggested chain of events in normal, tetraplegic and 'Shy-Drager' subjects on tilting in relation to the state of the afferent arteriole in the kidney and the release of renin. The tetraplegic subjects had transection of the cord in the high cervical region, while patients with the 'Shy-Drager' syndrome had autonomic neuropathy particularly associated with degeneration of the anteromedial tracts in the upper spinal cord and brain stem.

densa, and it is not impossible even in the experiments of *Thurau et al.* (48) and *Thurau and Schnermann* (49) with alterations in tubular fluid composition, that some more direct effect could be taking place around the arteriole. It is therefore necessary to look for what might be termed the most reasonable interpretation of the evidence, since very direct approaches to the juxtaglomerular cells are few, a tribute to the difficulty of the experiments.

I think the most reasonable interpretation of the events in the afferent arteriole and juxtaglomerular cells is that renin is released when the afferent arteriole dilates, and is therefore seen under all circumstances when, with either reduction of arterial pressure or of renal artery pressure directly, the arteriole dilates in order to maintain constancy of renal blood flow; in other words, autoregulation of flow (fig. 1). This is supported by those experiments usually in anaesthetised animals in which reduction of arterial pressure in the renal artery leads to renin release while flow is constant and is not increased further when the pressure is low enough to cause reduction of renal blood flow (14, 20). In conscious man on tilting or standing, renin is released very rapidly (32) and this is not associated with a fall of arterial pressure, which fits well with the concept of sympathetic stimulation with the release of small amounts of noradrenaline in the neighbourhood of the juxtaglomerular cells. In normal subjects, this renin release is blocked by propranolol (32) and other β-blockers, and the final receptor within the kidney seems undoubtedly to be of the β variety. The normal situation may be contrasted with that seen in patients with transected spinal cords in the high cervical region who are cut off from cerebral sympathetic centres. They elevate their renin levels on tilting very rapidly and to higher levels than in normal subjects (30). They also lower their pressure under these circumstances, and I believe it is most reasonable to suggest that renin release is due to autoregulation within the kidney which must be associated with dilatation of the afferent arteriole (20). This release cannot be blocked by propranolol (*Mathias, C.J., Dulieu, J., Peart, W.S. and Tunbridge, R.D.G.,* unpublished observations) and I think is quite analogous to the experiments in the anaesthetised animal where renal artery pressure is directly reduced leading to similar autoregulation for preservation of renal flow.

α versus β Stimulation

From a pharmacological point of view, a strong case can be made for considering that α stimulation and vasoconstrictors in general inhibit renin release (37, 55, 56, 58), and this is particularly shown in isolated perfused kidneys, but also in whole animal experiments. This can be contrasted with the ready release of renin by β-stimulators (2, 51, 57). It can be shown that α-stimulators like methoxamine, will block the release of renin induced by β-stimulators like isoprenaline (56). How then does noradrenaline, which is released at sympathetic nerve endings, fit into this general concept where it is supposed that sympathetic stimulation is the most important factor in acute release of renin under many physiological circumstances? Noradrenaline is both an α- and a β-stimulator, and from experiments in the isolated perfused kidney, there seems little doubt that the quantitative aspects are important in that in the very small dose range, the β effect predominates, making noradrenaline a very potent renin releaser, but that as the dose is increased, the α effect comes to dominate and renin release is reduced (54). It is therefore only necessary to propose that sympathetic stimulation, say on standing in normal man, leads to release of very small amounts of noradrenaline in the vicinity of the juxtaglomerular cells, thus causing renin release, and that under most circumstances the α effect is not seen.

Vasoconstrictors versus Vasodilators

Since the inhibitory effects of angiotensin in all experimental situations have been well demonstrated (9, 52, 58), and since the inhibitory effect of the final product of an enzyme substrate reaction is a very natural controlling

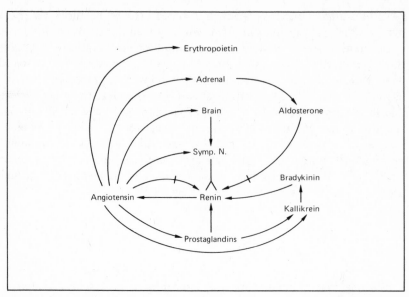

Fig. 2. Some but not all of the known physiological and pharmacological relationships of substances released within the kidney and reaching the blood, extracellular fluid, lymph and urine. A bar across the line refers to an inhibitory pathway.

mechanism, it is necessary to consider whether angiotensin is in a special category. I suggest that it is not and that most substances that cause intra-renal vasoconstriction inhibit the release of renin. Methoxamine has already been mentioned and vasopressin is another (9, 53). I would therefore like to suggest that the real balance is not just between α- and β receptors, but between vasoconstrictors and vasodilators in relation to renin release. The latter group includes glucagon (57), bradykinin and prostaglandin E_2. Both bradykinin, through the action of kallikrein, and prostaglandins are liberated within the kidney to enter the urine, lymph or blood (15, 26, 29). The complex relationships possible here are shown by the observations that all the substances, including angiotensin, seem to be able to release each other (fig. 2), and because of their different vascular and other pharmacological effects, to modify the final result within the kidney (28). A clear example is the way in which the vasoconstrictor effect of angiotensin injected into the renal artery is enhanced by the prior administration of indomethacin, which suggests that the vasodilator prostaglandin E_2, reduced by indomethacin (12), is usually opposing the vasoconstriction of angiotensin (1). The effects on renin release could therefore be complicated and even sometimes apparently paradoxical if these possibilities are not well in mind.

Macula densa Hypothesis

This whole discussion therefore has concentrated on vascular changes in the region of the juxtaglomerular cells. Where does that leave the macula densa hypothesis (48, 49), which in its most recent form suggests that chloride flux at this region (43) leads to renin release and local action with substrate, perhaps within the arteriolar wall, causing constriction and reduction of flow to the particular nephron concerned. I find it difficult to incorporate the events which have to occur in relation to the macula densa hypothesis, since renin release seems particularly related to afferent arteriole vasodilatation produced in a wide variety of ways. The experiments with the non-filtering kidney first performed by *Blaine et al.* (5) and then again with a different model (42), all show that the vascular component will operate satisfactorily in the absence presumably of any changes in fluid opposite the macula densa in the distal convoluted tubule. This can never be used as evidence that it does not play a role, however, but my concern is with the relation of vasodilatation and vasoconstriction to renin release.

Final Common Pathway

If there is such a thing as a final common pathway, then the following proposals may be of importance. Since the sympathetic nervous system was invoked at an early stage, and since the relation of β-receptors to cyclic-AMP release within the cell was made prominent by the work of *Sutherland* (45) and *Sutherland and Robison* (46), efforts have been made to incorporate that pathway in renin release. The evidence seems unconvincing (31, 35, 47, 59, 60). In general work with isolated glomeruli and their attachments or kidney slices has proved very difficult, not only in its execution because of poor renin production by such isolated tissues (6, 17, 33), but by the possibility of release of many other substances, particularly from slices or homogenates, which are uncontrolled and make interpretation very difficult. For this reason some conflicting results have emerged, particularly about the pharmacological release of renin. I would therefore rely more on whole kidney perfusion *in vivo* or *in vitro* at present.

Calcium Flux Hypothesis

My colleagues and I, as a result of experiments with the isolated perfused kidney (22, 27, 36, 55), have proposed that renin release depends very heavily on calcium flux across the juxtaglomerular cell. We have based this concept on the derivation of the juxtaglomerular cell from smooth muscle cells (3), and the fact that it was likely that they would still share some of the receptors and some of the characteristics of smooth muscle cells. It is known, of course, that smooth muscle contraction depends heavily upon changes in net calcium flux (44), and in the case of many endocrine cells the secretory process has been linked to a calcium-triggered stimulus-secretion process (19). A series of experiments involving changes in external calcium concentration and the use of substances which interfere with calcium flux, such as lanthanum (27), EDTA and EGTA (36), provided very suggestive evidence in favour, but perhaps the strongest evidence has come from the use of ionophores. These are peptides which have the property of transferring ions from one side of the cell membrane to the other (38, 39), and in this case the clearest evidence has come from the use of the ionophore A23187 which has a selective power in transporting calcium and magnesium (10, 16, 40). The experiments show that the inhibition of renin release in the presence of the ionophore is markedly related to the presence of external calcium, and that the increase of renin release conversely is associated with reduction of external calcium (22). The most ready explanation is that a net inflow of calcium ions, thus increasing the internal calcium concentration of the juxtaglomerular cells, leads to inhibition of renin release, and that a reduction of the internal calcium leads to an increase of renin release (fig. 3). The parallel change in arteriolar smooth muscle is of course contraction on the one hand and relaxation on the other. It might be possible to extend this hypothesis to involve vasoconstrictors and vasodilators since on the whole vasoconstriction is associated with the necessity for freely available calcium, and in its absence it has been shown that, for example, angiotensin does not contract renal blood vessels, and further, does not inhibit renin release (55). It would then be necessary to postulate that drugs like isoprenaline or glucagon, which are vasodilators, increase renin release by increasing the net efflux of calcium ions. This is the hypothesis I wish to present as a final common pathway, but obviously further work is needed to prove or disprove the general idea.

It now seems possible within the kidney to join together the vascular, sympathetic and juxtaglomerular cell physiology in a way which explains most of the physiological happenings around renin release. What is much less certain is the way in which such obvious stimuli as sodium deprivation and aldosterone exert their effects. In contrast to most of the stimuli discussed above, they produce their effects slowly and I would stress that sodium deprivation produces increase of plasma renin activity as well as of aldosterone in man with relatively small negative sodium balances of the order of 200 mEq and without any change in plasma sodium levels (7). The final stimulus to the juxtaglomerular cell is quite unknown but propranolol has been shown to reduce the increased plasma renin activity induced by a low sodium diet (32). This, however, only proves that the sympathetic nervous system is still functioning in renin release under

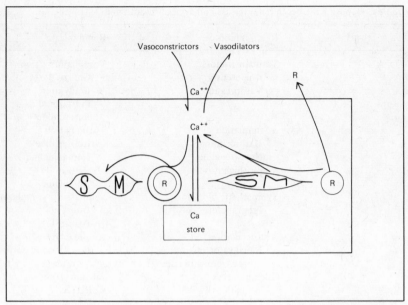

Fig. 3. Diagrammatic representation of the calcium flux hypothesis relating renin release (R) and arteriolar smooth muscle contraction (SM). On the left, influx of calcium leading to increased intracellular ionised calicum (Ca⁺⁺) causes, on the one hand, smooth muscle contraction, and on the other, inhibition of renin release. On the right, efflux of calcium leading to reduction of intracellular ionised calcium causes smooth muscle relaxation and increased release of renin. Further control of the level of intracellular ionised calcium is shown by the flux between intracellular storage sites (Ca store). On this hypothesis, the various factors known to affect either smooth muscle contraction or renin release, for example, vasoconstrictors versus vasodilators, as in the figure, may require interpretation in terms of net calcium flux (see table I).

these circumstances and the final levels achieved were still higher than in the same subjects on a 100 mEq sodium diet. The inhibition of renin release and reduction of renal renin content by aldosterone is also of uncertain nature, and the inhibition may be reversed by spironolactone (8). In patients with primary hyperaldosteronism, the plasma renin activity takes several days or weeks to return to normal with either large doses of spironolactone or removal of a Conn's tumour (11, 25). A direct effect on synthesis is likely, and factors controlling synthesis as opposed to release need much more study.

Finally, my view on the mechanisms involved in renin release may be summarised by considering the opposing pairs of factors concerned in both renin release and arterial smooth muscle contraction (table I).

Table I. Opposing factors in renin release and arterial smooth muscle contraction

	Renin inhibition	Renin release
Quick	Vasoconstrictors	Vasodilators
	Angiotensin	Glucagon
	Vasopressin	Bradykinin
		Prostaglandins
		Furosemide
	α-Stimulation	β-Stimulation
	Methoxamine	Isoprenaline
	Noradrenaline	Noradrenaline
	(high dose)	(low dose)
		Adrenaline
	Sympathetic blockade	Sympathetic stimulation
	Propranolol	Direct
		Posture
	Autoregulation	Autoregulation
	Renal vasoconstriction	Renal vasodilatation
	after pressure rise	after pressure fall
	Calcium influx	Calcium efflux
	Ionophores	EDTA
		Ionophores
Slow	Mineralocorticoids	Mineralocorticoid block
	Aldosterone	Spironolactone
	Sodium load	Sodium deprivation
		Diuretics
		Thiazide

Summary

Renin release is believed to depend more on vasodilatation in the afferent arteriole than on any other factor. This allows for opposing effects of vasoconstriction and vasodilatation produced by sympathetic nerve stimulation by drugs or by autoregulation, to be interpreted in relation to the study of stretch of the afferent arteriole. This reduces but does not remove the necessity for alternate control through the macula densa. A final common pathway for all these stimuli is suggested through alterations in net calcium flux in the juxtaglomerular cell where increased intracellular calcium inhibits, and decreased intracellular calcium increases, renin release.

References

1 *Aiken, J.W. and Vane, J.R.:* Intrarenal prostaglandin release attenuates the renal vasoconstrictor activity of angiotensin. J. Pharmac. exp. Ther. *184:* 678–687 (1973).

2 *Assaykeen, T.A.; Clayton, P.L.; Goldfien, A., and Ganong, W.F.:* Effect of alpha- and beta-adrenergic blocking agents on the renin response to hypoglycaemia and epinephrine in dogs. Endocrinology *87:* 1318–1322 (1970).

3 *Barajas, L. and Latta, H.:* Structure of the juxtaglomerular apparatus. Circulation Res. *20/21:* suppl. II pp. 15–28 (1967).

4 *Barajas, L. and Müller, J.:* The innervation of the juxtaglomerular apparatus and surrounding tubules. A quantitative analysis by serial section electron microscopy. J. Ultrastruct. Res. *43:* 107–132 (1973).

5 *Blaine, E.H.; Davis, J.O., and Prewitt, R.L.:* Evidence for a renal vascular receptor in control of renin secretion. Am. J. Physiol. *220:* 1593–1597 (1971).

6 *Blendstrup, K.; Leyssac, P.P.; Poulsen, K., and Skinner, S.L.:* Characteristics of renin release from isolated superfused glomeruli *in vitro.* J. Physiol., Lond. *246:* 653–672 (1975).

7 *Boyd, G.W.; Adamson, A.R.; Arnold, M.; James, V.H.T., and Peart, W.S.:* The role of angiotensin II in the control of aldosterone in man. Clin. Sci. *42:* 91–104 (1972).

8 *Brown, J.J.; Davies, D.L.; Lever, A.F.; Peart, W.S., and Robertson, J.I.S.:* Plasma concentration of renin in a patient with Conn's syndrome with fibrinoid lesions of the renal arterioles. The effect of treatment with spironolactone. J. Endocr. *33:* 279–293 (1965).

9 *Bunag, R.D.; Page, I.H., and McCubbin, J.W.:* Inhibition of renin release by vasopressin and angiotensin. Cardiovasc. Res. *1:* 67–73 (1967).

10 *Cashwell, A.H. and Pressman, B.C.:* Kinetics of transport of divalent cations across sarcoplasmic reticulum vesicles induced by ionophores. Biochem. biophys. Res. Commun. *49:* 292–300 (1972).

11 *Conn, J.W.; Cohen, E.L., and Rovner, D.R.:* Suppression of plasma renin activity in primary aldosteronism. J. Am. med. Ass. *190:* 213–221 (1964).

12 *Davis, H.A. and Horton, E.W.:* Output of prostaglandins from the rabbit kidney, its increase on renal nerve stimulation and its inhibition by indomethacin. Br. J. Pharmacol. *46:* 658–675 (1972).

13 *Davis, J.O.:* The regulation of renin release; in *Onesti, Kim and Moyer* Hypertension: mechanisms and management, pp. 617–629 (Grune & Stratton, New York 1973).

14 *Davis, J.O.:* The control of renin release. Am. J. Med. *55:* 333–350 (1973).

15 *De Bono, E. and Mills, I.H.:* Simultaneous increases in kallikrein in renal lymph and urine during saline infusion. J. Physiol., Lond. *241:* 127–128P (1974).

16 *Desmedt, J.E. and Hainaut, K.:* The effect of A23187 ionophore on calcium movements and contraction processes in single barnacle muscle fibres. J. Physiol., Lond. *257:* 87–107 (1976).

17 *De Vito, E.; Gordon, S.B.; Cabrera, R.R., and Fasciolo, J.C.:* Release of renin by rat kidney slices. Am. J. Physiol. *219:* 1036–1041 (1970).

18 *Doležel, S.; Edvinsson, L.; Owman, C., and Owman, T.:* Fluorescence histochemistry and autoradiography of adrenergic nerves in the renal juxtaglomerular complex of mammals and man, with special regard to the efferent arteriole. Cell Tissue Res. *169:* 211–220 (1976).

19 *Douglas, W.W.:* Involvement of calcium in exocytosis and the exocytosis-vesiculation sequence. Biochem. Soc. Symp. *39:* 1–28 (1974).

20 *Eide, I.; Løyning, E., and Kiil, F.:* Evidence for hemodynamic autoregulation of renin release. Circulation Res. *32:* 237–245 (1973).

21 *Faarup, P.:* Morphological aspects of the renin-angiotensin system (Copenhagen 1971).

22 *Fynn, M.; Onomakpome, N., and Peart, W.S.:* The effects of ionophores (A23187 and

R02-2985) on renin secretion and renal vasoconstriction. Proc. R. Soc. Lond. B. *199:* 199–212 (1977).

23 *Goormaghtigh, N.:* Les segments neuro-myo-artériels juxtaglomérulaires du rein. Archs Biol., Paris *43:* 575–591 (1932).

24 *Goormaghtigh, N.:* Facts in favour of an endocrine function of the renal arterioles. J. Path. Bact. *57:* 392–393 (1945).

25 *Kirkendall, W.M.; Fitz, A., and Armstrong, M.L.:* Hypokalaemia and the diagnosis of hypertension. Dis. Chest *45:* 337–345 (1964).

26 *Lee, J.B.; Covino, B.G.; Takman, B.H., and Smith, E.R.:* Renomedullary vasodepressor substance, medullin. Isolation, chemical characterization and physiological properties. Circulation Res. *17:* 57–77 (1965).

27 *Logan, A.G.; Tenyi, I.; Peart, W.S.; Breathnach, A.S., and Martin, B.G.H.:* The effect of lanthanum on renin secretion and renal vasoconstriction. Proc. R. Soc. Lond. B *195:* 327–342 (1977).

28 *McGiff, J.C.; Crowshaw, K.; Terragno, N.A., and Lonigro, A.J.:* Renal prostaglandins: possible regulators of the renal actions of pressor hormones. Nature, Lond. *227:* 1255–1257 (1970).

29 *McGiff, J.C.; Terragno, N.A.; Malik, K.U., and Lonigro, A.J.:* Release of a prostaglandin E-like substance from canine kidney by bradykinin. Circulation Res. *31:* 36–43 (1972).

30 *Mathias, C.J.; Christensen, N.J., Corbett, J.L.; Frankel, H.L.; Goodwin, T.J., and Peart, W.S.:* Plasma catecholamines, plasma renin activity and plasma aldosterone in tetraplegic man, horizontal and tilted. Clin. Sci. molec. Med. *49:* 291–299 (1975).

31 *Michelakis, A.M.; Caudle, J., and Liddle, G.W.: In vitro* stimulation of renin production by epinephrine, norepinephrine and cyclic AMP. Proc. Soc. exp. Biol. Med. *130:* 748–753 (1969).

32 *Michelakis, A.M. and McAllister, R.G.:* The effect of chronic adrenergic receptor blockade on plasma renin activity in man. J. clin. Endocr. Metab. *34:* 386–394 (1972).

33 *Morris, B.J.; Nixon, R.L., and Johnston, C.I.:* Release of renin from glomeruli isolated from rat kidney. Clin. exp. Pharmacol. Physiol. *3:* 37–47 (1976).

34 *Nilsson, O.:* The adrenergic innervation of the kidney. Lab. Invest. *14:* 1392–1395 (1965).

35 *Peart, W.S.; Quesada, T., and Tenyi, I.:* The effects of cyclic adenosine 3',5'-monophosphate and guanosine 3',5'-monophosphate and theophylline on renin secretion in the isolated perfused kidney of the rat. Br. J. Pharmacol. *54:* 55–60 (1975).

36 *Peart, W.S.; Quesada, T., and Tenyi, I.:* The effects of EDTA and EGTA on renin secretion. Br. J. Pharmacol. *59:* 247–252 (1977).

37 *Pettinger, W.A.; Keeton, T.K.; Campbell, W.B., and Harper, D.C.:* Evidence for a renal α-adrenergic receptor inhibiting renin release. Circulation Res. *38:* 338–346 (1976).

38 *Pressman, B.C.:* Properties of ionophores with broad range cation selectivity. Fed. Proc. Fed. Am. Socs exp. Biol. *32:* 1693–1703 (1973).

39 *Pressman, B.C.; Harris, E.J.; Jagger, W.S., and Johnson, J.H.:* Antibiotic-mediated transport of alkali ions across lipid barriers. Proc. natn. Acad. Sci. USA *58:* 1949–1956 (1967).

40 *Reed, P.W. and Lardy, H.A.:* A23187: a divalent cation ionophore. J. biol. Chem. *247:* 6970–6077 (1972).

41 *Ruyter, J.H.C.:* Über einen merkwürdigen Abschnitt der Vasa afferentia in der Mäuseniere. Z. Zellforsch. mikrosk. Anat. *2:* 242–248 (1925).

42 *Sadowski, J. and Wocial, B.:* Renin release and autoregulation of blood flow in a new

model of non-filtering non-transporting kidney. J. Physiol., Lond. *266:* 219–233 (1977).

43 *Schnermann, J.:* Regulation of filtrate formation by feedback. Proc. 6th Int. Congr. Nephrol., Florence 1975, pp. 230–234 (Karger, Basel 1976).

44 *Somlyo, A.P. and Somlyo, A.V.:* Vascular smooth muscle. I. Normal structure, pathology, biochemistry and biophysics. Pharmac. Rev. *20:* 197–272 (1968).

45 *Sutherland, E.W.:* Studies on the mechanism of hormone action. Science *177:* 401–408 (1972).

46 *Sutherland, E.W. and Robison, G.A.:* Role of cyclic 3'5' AMP in response to catecholamines and other hormones. Pharmac. Rev. *18:* 145–161 (1966).

47 *Tagawa, H. and Vander, A.J.:* Effects of adenosine compounds on renal function and renin secretion in dogs. Circulation Res. *26:* 327–338 (1970).

48 *Thurau, K.; Dahlheim, H.; Grüner, A.; Mason, J., and Granger, P.:* Activation of renin in the single juxtaglomerular apparatus by sodium chloride in the tubular fluid at the macula densa. Circulation Res. *30/31:* suppl. II, pp. 182–186 (1972).

49 *Thurau, K. und Schnermann, J.:* Die Natriumkonzentration an den Macula-densa-Zellen als regulierender Faktor für das Glomerulumfiltrat. Klin. Wschr. *43:* 410–413 (1965).

50 *Tobian, L.:* Interrelationship of electrolytes, juxtaglomerular cells and hypertension. Physiol. Rev. *40:* 280–312 (1960).

51 *Ueda, H.; Yasuda, H.; Takabatake, Y.; Iizuka, M.; Iizuka, T.; Ihori, M., and Sakamoto, Y.:* Observations on the mechanism of renin release by catecholamines. Circulation Res. *26/27:* suppl. II, pp. 195–200 (1970).

52 *Vander, A.J. and Geelhoed, G.W.:* Inhibition of renin secretion by angiotensin II. Proc. Soc. exp. Biol. Med. *120:* 399–403 (1965).

53 *Vandongen, R.:* Inhibition of renin secretion in the isolated rat kidney by antidiuretic hormone. Clin. Sci. molec. Med. *49:* 73–76 (1975).

54 *Vandongen, R. and Greenwood, D.M.:* The stimulation of renin secretion by non-vasoconstrictor infusions of adrenaline and noradrenaline in the isolated rat kidney. Clin. Sci. molec. Med. *49:* 609–612 (1975).

55 *Vandongen, R. and Peart, W.S.:* Calcium dependence of the inhibitory effect of angiotensin on renin secretion in the isolated perfused kidney of the rat. Br. J. Pharmacol. *50:* 125–129 (1974).

56 *Vandongen, R. and Peart, W.S.:* The inhibition of renin secretion by alpha-adrenergic stimulation of the isolated rat kidney. Clin. Sci. molec. Med. *47:* 471–479 (1974).

57 *Vandongen, R.; Peart, W.S., and Boyd, G.W.:* Adrenergic stimulation of renin secretion in the isolated perfused rat kidney. Circulation Res. *32:* 290–296 (1973).

58 *Vandongen, R.; Peart, W.S., and Boyd, G.W.:* Effect of angiotensin II and its nonpressor derivatives on renin secretion. Am. J. Physiol. *226:* 277–282 (1974).

59 *Winer, N.; Chokshi, D.S., and Walkenhorst, W.G.:* Effect of cyclic AMP, sympathomimetic amines and adrenergic receptor antagonist on renin secretion. Circulation Res. *29:* 239–248 (1971).

60 *Yamamoto, K.; Okahara, T.; Abe, Y.; Ueda, J.; Kishimoto, T., and Morimoto, S.:* Effects of cyclic AMP and dibutyryl cyclic AMP on renin release *in vivo* and *in vitro.* Jap. Circul. J. *37:* 1271–1276 (1973).

W.S. Peart, MD, FRCP, FRS, Professor of Medicine, Medical Unit, St. Mary's Hospital, *London W2* (England)

Contr. Nephrol., vol. 12, pp. 16–26 (Karger, Basel 1978)

What Makes the Renin-Angiotensin System a Pathogenic Factor?

R. Dietz, H. Haebara and F. Gross

Department of Pharmacology, University of Heidelberg, Heidelberg

The renin-angiotensin system (RAS) is closely related to the regulation of sodium balance, both in the normotensive and in the hypertensive state. Angiotensin II (A II) is the most powerful humoral stimulant of aldosterone release (*Brown et al.,* 1973), although it has been shown that the secretion of aldosterone is controlled also in the absence of a functioning RAS (*Blair-West et al.,* 1972, 1973). Normally, the stimulation of the RAS, induced by negative sodium balance, does not cause a rise in blood pressure. However, when renin is released in amounts that do not correspond to the state of sodium balance, the increased formation of A II may provoke high blood pressure. One such situation may be renal artery stenosis, which induces a marked increase in renin release, provided that the contralateral kidney remains untouched. Today, it is generally accepted that the RAS plays a significant part in the pathogenesis of this type of renal hypertension. However, after unilateral nephrectomy, renal artery stenosis is not followed by an activation of the RAS, although the blood pressure rises to even higher values than in the presence of an intact contralateral kidney.

It has been suggested that the activation of the RAS after unilateral renal artery stenosis is the consequence of a negative sodium balance, which is caused by salt and fluid loss through the intact contralateral kidney that is exposed to the high blood pressure. If this were the case, either removal of this kidney or stenosis of its artery should prevent pressure-induced saluresis (*Gross et al.,* 1975). Furthermore, it should also be possible to inhibit the activation of the RAS despite the presence of the contralateral kidney by administering a mineralocorticoid and salt after the clip has been placed on one renal artery.

The studies reported in this paper have been planned to elucidate the significance of the RAS in the pathogenesis of various types of experimental renal hypertension. Three types of renal artery stenosis have been compared with one another: type I − stenosis of one renal artery in the presence of an intact contralateral kidney ('two-kidney hypertension' of other investigators); type II − renal artery stenosis of the kidney remaining after unilateral nephrectomy ('one-kidney hypertension'); type III − bilateral renal artery stenosis.

Material and Methods

Male Sprague-Dawley rats (SIV-50 strain, Ivanovas, Kisslegg, Allgäu, FRG) of an average weight of 150 g were used. After an adaptation period of 4−5 days, all animals were kept in individual cages and had free access to food (ssniff®) and demineralized water.

Various Types of Renal Artery Stenosis

All animals were operated under light ether anesthesia. The following groups were formed: (1) Sham-operated controls (n = 20); after a dorsal midline incision, both renal arteries were exposed. (2) Type I (n = 25); after dorsal exposure of the left kidney, a silver clip, internal diameter 0.2−0.22 mm, was placed on the renal artery; the right kidney remained untouched. (3) Type II (n = 20); after removal of the right kidney, the left renal artery was constricted by a silver clip, internal diameter 0.2−0.22 mm. (4) Type III (n = 20); silver clips of the same size as in the other experiments were placed on both renal arteries.

Suppression of the RAS by DOC and Salt

Rats of type I renal hypertension (n = 14) received 5 mg/kg desoxycorticosterone-acetate (DOCA) in oily solution twice daily, beginning 4 days before the clamping of the renal artery and continuing for 20 days. Drinking fluids: 2% saline solution and water by free choice.

In another series (n = 53), 3- to 4-week-old rats of an average weight of 40−50 g received injections of 10 mg/kg desoxycorticosterone-trimethylacetate (DOC-TMA) twice weekly. Unilateral renal artery stenosis was induced after 4 weeks (n = 43); a further group was sham-operated (n = 10) at the same time. After the operation, the animals were followed up for another 2 weeks, still receiving DOC-TMA. During the whole experiment of 6 weeks' duration, the drinking fluid was 1% saline.

Genetic Hypertension

Rats with spontaneous hypertension, the stroke-prone substrain of the Okamoto-Aoki strain, were used. Starting from the age of 7 weeks, three groups of rats received a diet that differed in its sodium content for a period of 20 days. The first group (n = 10) had a low-sodium diet, containing less than 1 mmol sodium/kg. The second group (n = 10) had a diet with normal sodium content of 100 mmol sodium/kg. The third group (n = 10) received a diet with ten times the normal sodium content, i.e. 1,000 mmol sodium/kg chow.

In all animals, the blood pressure was measured by means of tail plethysmography under light ether anesthesia twice weekly. A II in plasma was determined by direct radioimmunoassay of A II (Oster et al., 1973). Plasma concentrations of aldosterone and of corticosterone were estimated by radioimmunoassay (Vecsei, 1974).

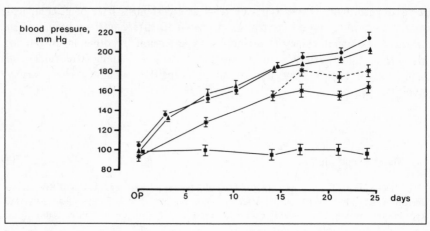

Fig. 1. Increase in blood pressure in three types of renal artery stenosis. Type I, n = 25 (*) type II, n = 20 (▲); type II, n = 20 (●). In the rats with type I hypertension, the broken line indicates the blood pressure of the animals with signs of acute sodium loss ('malignant' phase). Control animals, n = 20 (■).

Results

Renal Artery Stenosis

After reduction of renal blood flow, the blood pressure rose promptly in all three groups, but the increase was more rapid and more pronounced in the rats with type II and type III renal artery stenosis than in those with type I (fig. 1). In the latter group, two courses of hypertension could be distinguished. One part of the animals developed symptoms of negative sodium and fluid balance, leading to an acute syndrome of salt loss. A marked stimulation of the RAS resulted, during which the blood pressure values were higher than in the animals that showed no signs of negative sodium balance. In former studies, we had characterized these two courses of hypertension as 'benign' and 'malignant' (*Dauda et al.,* 1973; *Möhring et al.,* 1975). The plasma concentration of A II was highest in the rats with type I renal artery stenosis which developed a syndrome of acute sodium loss, whereas it was about twice normal in the animals without signs of acute sodium loss. In the rats with renal artery stenosis of type II and type III, the plasma concentration of A II was in the upper normal range (fig. 2). The changes in the plasma aldosterone concentration were similar to those in the plasma A II concentration; a slight increase was seen in the rats with type II stenosis, whereas it remained in the normal range in type III hypertension. Plasma corticosterone was elevated in both forms of type I hypertension and most markedly increased in type III hypertension.

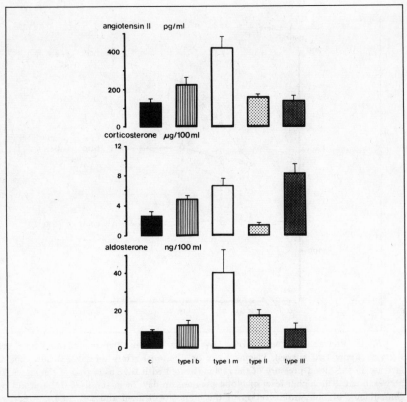

Fig. 2. Plasma concentration of angiotensin II (pg/ml), of corticosterone (μg/100 ml), and of aldosterone (ng/100 ml) in various types of renal artery stenosis. c = Controls; type I = unilateral stenosis, contralateral kidney intact; Ib = 'benign' course of hypertension; Im = 'malignant' course of hypertension; type II = unilateral stenosis after removal of the contralateral kidney; type III = stenosis of both renal arteries. Same animals as in figure 1 at the end of the experiment (day 25).

Renal Artery Stenosis and Administration of Mineralocorticoids

To find out whether the activation of the RAS is of significance for the development of type I hypertension, two experiments were carried out in which the RAS was suppressed by the administration of DOC. When DOCA was injected daily from the time the clip was placed on one renal artery and a 2% saline solution was given as drinking fluid in addition to water, the blood pressure rose similarly as in the rats that had only water to drink and did not receive the mineralocorticoid. However, while, in the latter group, the concentration of A II in plasma was about four times normal, it remained in the normal range in the animals given DOC and saline. Hence, in rats with unilateral renal

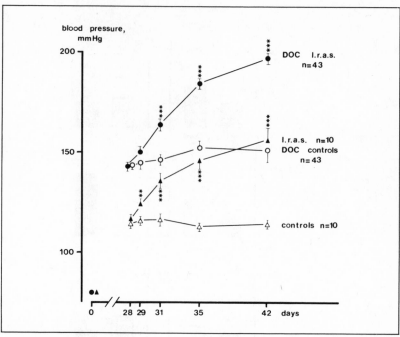

Fig. 3. Pretreatment with DOC-TMA during 4 weeks before unilateral renal artery stenosis (upper two curves). Placing a clip on one renal artery on day 28 causes the same increase in the blood pressure of the rats pretreated with DOC as in that of the rats without pretreatment. The higher level of blood pressure on day 28 in the DOC-TMA group is the consequence of the administration of the mineralocorticoid and salt. l.r.a.s. = Left renal artery stenosis.

artery stenosis and an intact contralateral kidney, the blood pressure may rise independently of the activity of the RAS. The administration of DOC and salt could not prevent the symptoms of acute sodium loss, indicated by loss of body weight, further elevation in water and saline intake, and an impaired general condition of the animal.

When young rats, at the age of 25 days, received DOC-TMA and a 1% saline solution as drinking fluid for 4 weeks, their blood pressure rose by about 25 mm Hg. The placing of a clip on one renal artery of these rats produced the same increase in blood pressure as in control rats that had not been pretreated with the mineralocorticoid and had no additional salt supply. A fortnight after the induction of the renal artery stenosis, the pressure was about 45 mm Hg higher in the animals that had received the mineralocorticoid, which corresponded approximately to the higher pressure on the day when the clip was placed on the renal artery (fig. 3). At the end of the experiment, the plasma concentration of

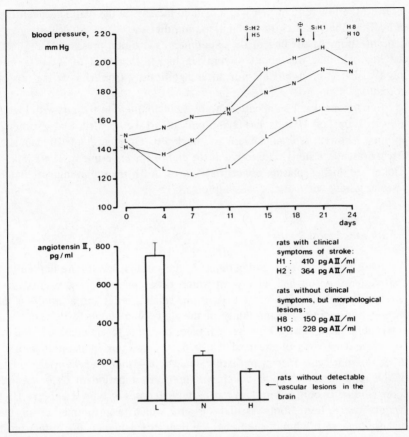

Fig. 4. Effect of high or low sodium intake on the blood pressure in spontaneously hypertensive rats of the stroke-prone substrain. Upper part: blood pressure of the rats which received a normal (N), a high (H), or a low (L) sodium diet. For all three groups, n = 10. Lower part: plasma angiotensin II concentrations in the three groups. In 5 rats with high sodium intake, symptoms of stroke occurred or cerebrovascular lesions were found.

A II was reduced in the rats that had received DOC, whereas it was increased in the controls which had no mineralocorticoid and only water to drink. The plasma concentration of aldosterone was markedly diminished in the steroid-treated group, but enhanced in the controls, whereas the plasma concentration of corticosterone was elevated in both groups.

Influence of Dietary Sodium
The effect of the stimulation or suppression of the RAS on the course of hypertension has been studied in spontaneously hypertensive rats. The rats that

received the low-sodium diet had a marked increase in the plasma concentration of A II, while the rats on the high-sodium diet showed a reduced plasma A II concentration. At the end of the experiment, the blood pressure of the rats on the high-salt diet was about 40 mm Hg higher than that of the rats on the low-salt diet (fig. 4), but did not differ significantly from that of the rats on a normal-salt diet.

Cerebrovascular complications were more frequent in the rats with high salt intake; 2 out of 10 rats had symptoms of stroke, 1 died, and 2 others, at autopsy, showed arterial damage in the brain. Only in 5 of 10 animals no cerebrovascular lesions were found at the end of the experiment. The 2 rats with stroke had higher plasma concentrations of A II than the animals that had vascular lesions without clinical symptoms.

Morphological Studies

In the rats with type I hypertension, the vascular lesions are limited to the contralateral, untouched kidney, in which thickening of the glomerular capillaries and of the intima and media of the small arteries and arterioles is seen. Initially, an increased permeability of the glomerular capillaries for large molecules (peroxidase, ferritin) is observed. Later on, more severe damage in the form of fibrinoid necrosis of the media may occur, resulting in nephrosclerosis. In other vascular beds, lesions are hardly detectable within 4 weeks after induction of the renal artery stenosis, which was the longest period of hypertension in those studies. In contrast to it, in renal artery stenoses of type II and type III, the kidneys were free from vascular lesions, which were numerous in other organs, such as the brain, the heart, the liver, the pancreas, the testes, and the mesentery. The most severe lesions were found in the unilaterally nephrectomized rats with stenosis of the remaining renal artery. Intima proliferation and marked thickening of the muscular layers of the media were observed in all vascular beds. The area of the mesenteric artery occasionally showed lesions similar to periarteritis nodosa, as has been described earlier by various authors (*Byrom,* 1969).

Discussion

In studies on experimental renal hypertension of rats, it is generally distinguished between the 'two-kidney' and the 'one-kidney' hypertension, the former being characterized by an intact contralateral kidney, whereas the latter refers to renal artery stenosis in the unilaterally nephrectomized animal. To these types I and II of renal hypertension, a third (type III) has to be added, which is also a 'two-kidney' hypertension, but with stenoses on both arteries. In earlier studies it has been stated that, in this type, slight differences in the degree of stenosis are unavoidable. Consequently, one kidney will always excrete slightly more water and solutes than the other. Furthermore, the renin content of the two kidneys will differ, since the kidney with the milder stenosis will have less renin than the kidney with the stronger reduction of renal blood flow (*Gross*, 1960).

The blood pressure increase in the three types of renal artery stenosis is similar, but higher values are reached in types II and III than in type I. The main difference of the three types is that in type I the intact contralateral kidney may become a sodium-losing kidney when the high blood pressure induces pressure diuresis and saluresis. The consequence will be a negative sodium and water balance, which results in an increased fluid intake and stimulates the RAS as well as the secretion of aldosterone. The syndrome of acute sodium loss develops, which is also characterized by reduced food intake and loss of body weight. This syndrome has been called 'acute malignant phase' of hypertension. Immediate relief from this syndrome could be obtained by offering a 0.9% saline solution as drinking fluid, which was avidly taken up by the animals, with the result that the whole condition improved and most of the monitored parameters turned towards normal values within 12–24 h (*Möhring et al.*, 1976b).

Provided that the sodium and fluid loss by the intact contralateral kidney is responsible for the acute salt-losing syndrome, removal of this kidney or protection from high blood pressure should have a preventive effect. Our experiments have demonstrated that this is the case, since neither the rats with type II nor those with type III renal artery stenosis showed symptoms of an acute phase of sodium loss with its clinical consequences, and, in particular, the RAS was not activated. On the other hand, the blood pressure increase was more marked and reached higher values in stenoses of types II and III than in type I. The discrepancy between the rise in blood pressure and the activity of the RAS in type II stenosis was observed years ago (*Regoli et al.*, 1962), but no corresponding data were available for bilateral stenosis (type III). It may be concluded that in these two types of renal artery stenosis the prevention of sodium and fluid loss is responsible for the fact that the stimulation of the RAS, which follows the stenosis of the renal artery, is not maintained. In additional experiments we observed, 24 h after placing the clip on one (type II) or on both (type III) renal arteries, a rise in the plasma concentration of A II, which, later on, was no longer demonstrable. It may be supposed that constriction of the renal artery causes a sudden fall of the renal perfusion pressure beyond the stenosis, which elicits a release of renin. However, because the poststenotic pressure increases continuously afterwards, the release of renin turns towards normal. In the presence of an intact contralateral kidney, the sodium and fluid loss, which occurs when the pressure rises, is the main stimulus for the activation of the RAS, which should serve the re-establishment of sodium balance by stimulating the secretion of aldosterone. Although such a mechanism may explain the variations in the activity of the RAS after different types of renal artery stenosis, it does not give a clue as to why the blood pressure rises in type II and type III renal artery stenosis.

However, also in hypertension of type I, the activity of the RAS is not simply correlated to the high blood pressure. Suppression of the RAS by DOC and salt does not at all change the development of hypertension after unilateral renal artery stenosis in the presence of an intact contralateral kidney. The consequence of the administration of the mineralocorticoid and salt is a higher incidence of cerebrovascular complications and of vascular lesions in various vascular beds. The additional salt supply could not completely prevent the acute sodium loss by the contralateral kidney and the corresponding symptoms, and in some rats, the plasma concentration of A II was about three to four times normal. It has been observed that, also in hypertension induced by DOC and salt, an acute malignant phase of hypertension, comparable to that described in type I hypertension, is seen in the presence of a completely suppressed plasma A II concentration (*Gavras et al.,* 1975). In our case, it may be assumed that the elevated plasma concentration of A II is an indicator of the situation in the sodium balance, but is of little pathogenic significance for the incidence of cerebrovascular lesions seen in the rats with type I hypertension, which received DOC and salt additionally.

The results of our experiments demonstrate that the severity of high blood pressure after renal artery stenosis cannot be correlated with the activity of the RAS. Suppression of the RAS by mineralocorticoid and salt even aggravated the vascular lesions and consequently the hypertensive complications. Other vaso-constrictor substances, such as the catecholamines or the antidiuretic hormone, have to be taken into consideration as hypertensinogenic mechanisms. The secretion of catecholamines or vasopressin is stimulated by loss of blood volume and may act as compensatory mechanism for the maintenance of blood pressure. In the case of the malignant phase of DOC hypertension, an increase in the secretion of antidiuretic hormone was demonstrated (*Möhring et al.,* 1976a), which, to a certain degree, may have replaced the vasoconstrictor action of angiotensin, completely abolished under these conditions. However, in the case of type II and type III renal artery stenosis, the secretion of antidiuretic hormone may even be reduced as a consequence of the tendency to fluid retention observed in these types of hypertension.

In conclusion, it may be stated that the RAS primarily serves the regulation of sodium and water balance and does do so independently of the height of blood pressure. In a relatively small number of cases of renal artery stenosis, a renin-dependent form of high blood pressure can be demonstrated, but in the majority the RAS does not play a significant part as a pathogenic factor in renal or other forms of hypertension.

Summary

Three types of renal hypertension in the rat have been compared with respect to blood pressure increase, activity of the RAS, and secretion of aldosterone and corticosterone: type I – unilateral stenosis of the renal artery in the presence of an intact contralateral kidney; type II – unilateral stenosis of the renal artery after contralateral nephrectomy; type III – bilateral stenosis of the renal arteries.

Blood pressure rose more rapidly and reached higher values in type II and type III hypertension than in type I hypertension. In the latter group, the activity of the RAS was more stimulated than in types II and III.

The marked stimulation of the RAS in type I hypertension is ascribed to the negative fluid and sodium balance, which is the consequence of a pressure-induced diuresis of the unclamped contralateral kidney.

Suppression of the activity of the RAS by a 4-week pretreatment with DOC-TMA and saline or by the administration of DOCA and saline as from the induction of renal artery stenosis did not prevent the development of hypertension caused by the clamping of one renal artery (type I).

In spontaneously hypertensive rats of the stroke-prone substrain, high dietary salt intake caused higher blood pressure values and a higher incidence of cerebral lesions than normal dietary salt intake. Low salt intake was followed by a marked stimulation of the RAS, but blood pressure rose only slightly and no symptoms of cerebrovascular lesions were observed.

It is concluded that, neither in hypertension induced by renal artery stenosis nor in spontaneously hypertensive rats, the RAS contributes significantly to the increase in blood pressure nor does it play a major part in the pathogenesis of vascular lesions. These seem to be related to the retention of sodium, which may be obtained by renal artery stenosis, by excessive salt intake, or by the administration of a mineralocorticoid and salt.

References

Blair-West, J.R.; Coghlan, J.P.; Cran, E.; Denton, D.A.; Funder, J.W., and Scoggins, B.A.: Contrived suppression of renin secretion during sodium depletion; in *Genest and Koiw* Hypertension – 1972, pp. 14–25 (Springer, Berlin 1972).

Blair-West, J.R.; Coghlan, J.P.; Cran, E.; Denton, D.A.; Funder, J.W., and Scoggins, B.A.: Increased aldosterone secretion during sodium depletion with inhibition of renin release. Am. J. Physiol. *224:* 1409–1414 (1973).

Brown, J.J.; Fraser, R.; Lever, A.F.; Morton, J.J.; Oelkers, W.; Robertson, J.I.S., and Young, J.: Further observations on the relationship between plasma angiotensin II and aldosterone during sodium deprivation; in *Sambhi* Mechanisms of hypertension. Proc. Int. Workshop Conf. Los Angeles 1973. Int. Congr. Series No. 302, pp. 148–154 (Excerpta Medica, Amsterdam 1973).

Byrom, F.B.: The hypertensive vascular crisis. An experimental study (Heinemann, London 1969).

Dauda, G.; Möhring, J.; Hofbauer, K.G.; Homsy, E.; Miksche, U.; Orth, H., and Gross, F.: The vicious circle in acute malignant hypertension of rats. Clin. Sci. molec. Med. *45:* suppl. 1, pp. 251–255 (1973).

Gavras, H.; Brunner, H.R.; Laragh, J.H.; Vaughan, E.D., jr.; Koss, M.; Cote, L.J., and Gavras, I.: Malignant hypertension resulting from deoxycorticosterone acetate and salt excess. Role of renin and sodium in vascular changes. Circulation Res. *36:* 300–309 (1975).

Gross, F.: Adrenocortical function and renal pressor mechanisms in experimental hypertension; in *Bock and Cottier* Essential hypertension. An international symposium, pp. 92–111 (Springer, Berlin 1960).

Gross, F.; Dietz, R.; Mast, G.J., and Szokol, M.: Salt loss as a possible mechanism eliciting an acute malignant phase in renal hypertensive rats. Clin. exp. Pharmacol. Physiol. *2:* 323–333 (1975).

Möhring, J.; Möhring, B.; Näumann, H.-J.; Philippi, A.; Homsy, E.; Orth, H.; Dauda, G.; Kazda, S., and Gross, F.: Salt and water balance, and renin activity in renal hypertension of rats. Am. J. Physiol. *228:* 1847–1855 (1975).

Möhring, J.; Möhring, B.; Petri, M., and Haack, D.: Is vasopressin involved in the pathogenesis of malignant desoxycorticosterone hypertension in rats? Lancet *i:* 170–173 (1976a).

Möhring, J.; Petri, M.; Szokol, M.; Haack, D., and Möhring, B.: Effects of saline drinking on malignant course of renal hypertension in rats. Am. J. Physiol. *230:* 849–857 (1976b).

Oster, P.; Hackenthal, E., and Hepp, R.: Radioimmunoassay of angiotensin II in rat plasma. Experientia *29:* 353–354 (1973).

Regoli, D.; Brunner, H.; Peters, G., and Gross, F.: Changes in renin content in kidneys of renal hypertensive rats. Proc. Soc. exp. Biol. Med. *109:* 142–145 (1962).

Vecsei, P.: Glucocorticoids: Cortisol, corticosterone, and compound S; in *Jaffe and Behrman* Methods of hormone radioimmunoassay, pp. 393–415 (Academic Press, New York 1974).

Prof. Dr. med. *F. Gross,* Department of Pharmacology, University of Heidelberg, Im Neuenheimer Feld 366, *D–6900 Heidelberg* (FRG)

Prostaglandins

Contr. Nephrol., vol. 12, pp. 27–40 (Karger, Basel 1978)

Prostaglandins and Renal Function

John C. McGiff, D. Alicia Terragno and Norberto A. Terragno

Department of Pharmacology, University of Tennessee, Center for the Health Sciences, Memphis, Tenn.

Introduction

The general conclusion that prostaglandins are primarily local or tissue hormones, which exert their effects at or near sites of synthesis, has particular significance for the kidney. The first functional studies on the capacity of the kidney to synthesize prostaglandins related this ability to modulation of the renal effects of angiotensin and norepinephrine (1–3). Release of prostaglandins during administration of pressor hormones resulted in attenuation of the renal vasoconstrictor and antidiuretic actions of the latter. Interactions of vasoactive hormones and prostaglandins are likely to occur primarily within the vascular wall as prostaglandin synthetase, or more precisely the cyclo-oxygenase, is present within the walls of blood vessels (4–6). This finding takes on added significance in view of recent studies concerning a prostaglandin mechanism which participates in the regulation of renin release (7, 8). As renin synthesis and storage within the kidney is intimately associated with vascular tissue (9), the presence of cyclo-oxygenase in the vascular wall fulfills the major criterion of local hormones, that is generation at sites in proximity to their major activity. It should be noted that these observations on prostaglandin mediated renin release have major implications for renal cortical sites of synthesis of prostaglandins which possibility was denied only several years ago (10) and later affirmed in a critical study (11) which demonstrated prostaglandin biosynthetic capacity for the renal cortex. The renal cellular element of greatest interest in the past relative to lipid metabolism has been the interstitial cell of the medulla which possesses the capacity to synthesize prostaglandins (12). The functional implications of the latter finding are that prostaglandins, after release into the medullary interstitium, may diffuse into either the blood or tubular fluid (13) and affect medullary blood flow (14) and, perhaps, transport of salt and water (15). In addition, it seems probable that cyclo-oxygenase is present within one or

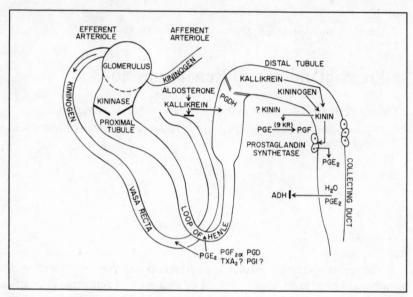

Fig. 1. The generation of kinins in the distal nephron and collecting ducts results in the release of prostaglandins which inhibit the effect of ADH and, thereby, participate in the excretion of solute-free water.

more cellular elements lining the urinary compartment, particularly the distal nephron and collecting ducts (16). This anatomical arrangement favors interaction of prostaglandins with ADH whereby the generation of prostaglandins contributes to the excretion of solute free water (17), at least in part due to inhibiting the action of ADH (18). In view of the widespread distribution of the cyclo-oxygenase amongst various cellular elements of the kidney, it is important to recognize that compartmentalization of prostaglandin metabolism may have important consequences for renal function. Thus, the effects consequent to generation of prostaglandins within the vasculature may be primarily restricted to that compartment although changes in renal medullary blood flow may have indirect effects on excretion of salt and water. Similarly, the functional consequences of interactions of the renal kallikrein-kinin and prostaglandin systems may be restricted primarily to the urinary compartment. Thus, entry of kallikrein into the distal tubules (19) and subsequent formation of kinin may result in a kinin mediated generation of prostaglandins (20) in the distal nephron and collecting ducts which interaction eventuates in the excretion of solute free water (fig. 1). This concept of segregation of the cyclo-oxygenases to, in the broadest sense, vascular, interstitial, and urinary compartments leads to consideration of two important corollaries. First, the cyclo-oxygenases associated with different cellular elements, be they tubular, interstitial, or vascular, may generate different products such as prostacyclin (PGI_2) by vascular tissue (21). The latter has been identified with antithrombotic and vasodilator activity of endothelial cells. PGE_2, which was advanced as the principal product of the cyclo-oxygenase of the kidney (22), is a potent antagonist of angiotensin II (1) and adrenergic nervous activity (23) and probably functions as the mediator of the prostaglandin mechanism which increases renal medullary blood flow in response to diverse stimuli (24). The second corollary is that the products of the cyclo-oxygenase in these various compartments may vary with experimental conditions as well as in health and in disease. Thus, thromboxane, a powerful vasoconstrictor product of cyclo-oxygenase (25), appears to be synthesized in negligible quantities in the normal kidney (26). However, major disturbances of renal function as evoked by acute ureteral ligation may result in significant amounts of thromboxane being generated intrarenally, presumably within the vascular compartment (26). As thromboxane is a powerful vasoconstrictor, it may contribute to the late increase in renal vascular resistance in response to ureteral obstruction (27).

As many studies on the possible contribution of prostaglandins to renal function have made use of aspirin-like compounds to inhibit cyclo-oxygenase (28), consideration will be given first to the variable effects of these agents on the kidney. Inhibition of renal cyclo-oxygenase by indomethacin and other aspirin-like compounds not only inhibits generation of PGE_2 and $PGF_{2\alpha}$, but thromboxane and prostacyclin synthesis as well. Therefore, the effects on renal function of aspirin-like drugs cannot be ascribed to suppression of any one product of cyclo-oxygenase. Generation of PGE_2, thromboxane, and prostacyclins from the common intermediate endoperoxides, PGG_2 and PGH_2, is enzymically regulated. Within the past year agents have been identified which inhibit with some selectivity thromboxane synthetase (29) as well as the enzyme which results in the formation of prostacyclin (30). Agents which inhibit selectively these enzymes are important for definition of the effects on renal function of the diverse products of cyclo-oxygenase.

Effects of Cyclo-Oxygenase Inhibitors on the Renal Circulation

The renal circulatory effects of inhibitors of cyclo-oxygenase (28) were shown by *Terragno et al.* (31) to vary with experimental conditions. This study demonstrated that in the acutely stressed dog the renal circulation is supported by a major prostaglandin component, withdrawal of which results in a decline in renal blood flow. Indomethacin does not affect renal blood flow in the anesthetized nonlaparotomized dog. Laparotomy performed on the chloralose-anesthetized dog induces a large increase in renal prostaglandin efflux associated with

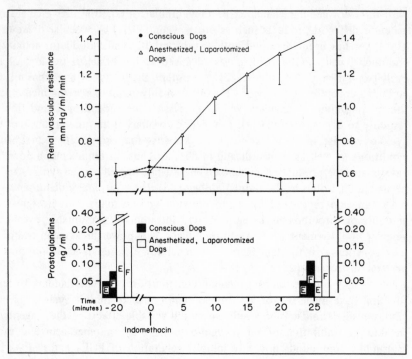

Fig. 2. The differential effect of indomethacin in anesthetized-laparotomized and in conscious dogs, on renal vascular resistance (upper) and on concentrations of 'PGE' and 'PGF' in renal venous blood (lower) at various time intervals. For renal vascular resistance, the vertical lines represent the standard error of the mean.

major circulatory effects of administered indomethacin. In the conscious dog prostaglandin concentrations in renal venous blood are considerably less than those in the chloralose-anesthetized-laparotomized dog and indomethacin affects neither renal vascular resistance nor the levels of prostaglandins, even in doses having major toxic effects. Indomethacin had been reported not to affect renal blood flow in the conscious animal (32, 33). In contrast, doses of indomethacin one-fifth those which were without effect in the conscious dog, when given to the anesthetized-laparotomized dog, effect large increases in renal vascular resistance, associated with reduction of renal venous prostaglandin concentrations to those levels observed in the conscious dog (fig. 2). Thus, there appears to be a component of renal prostaglandins which resists even toxic doses of indomethacin (10 mg/kg), sufficient to elicit bloody diarrhea. The significance of this component, although now undetermined, invites reconsideration of the possible contribution of prostaglandins, particularly those of renal cortical origin (11), to resting renal blood flow. Inasmuch as the inhibitory effect of aspirin-like drugs on prostaglandin synthesis may vary with either experimental conditions or the hormonal background (34), the inability of indomethacin to decrease renal blood flow in unanesthetized dogs may reflect the operation of one of these factors. Indeed, another aspirin-like compound, meclofenamic acid, reduced renal blood flow in conscious dogs when indomethacin was without effect (33). Further, after acute blood loss aspirin affects renal vascular resistance differently from indomethacin (35). These differences among the nonsteroidal antiinflammatory compounds may signify important differential effects of these drugs on prostaglandin metabolizing enzymes (36) including the cyclo-oxygenase which could contribute to their effects on renal blood flow. Moreover, the possibility of species-dependent effects of indomethacin should be considered; e.g., in the conscious rabbit indomethacin decreases renal blood flow (37).

Although the significance of a possible prostaglandin-dependent component of renal blood flow in the conscious resting dog remains to be resolved, it is clear that acute stress, such as hemorrhagic hypotension (38) or laparotomy (39), evokes a severalfold increase in the production of renal prostaglandins above 'basal levels'. Changes in synthesis of prostaglandins by the kidney not only affect renal blood flow but also its distribution intrarenally, particularly that fraction to the inner cortex and medulla. The evidence for this was first obtained in the isolated blood-perfused kidney of the dog by *Itskovitz et al.* (40) and has since been demonstrated in a number of experimental preparations (41, 42) including the conscious rabbit (37). The renal circulatory effects of enhanced prostaglandin synthesis in the acutely stressed dog are readily revealed by administration of an aspirin-like compound which results in a large reduction of renal blood flow, primarily that component to the inner cortex and medulla (24, 35, 42). It should be recalled that the zonal distribution of prostaglandin synthetase within the kidney (11) is opposite to that of renin (43), the highest activity being noted in the papilla and inner medulla, the least in the renal cortex. This anatomical arrangement in large part determines the effects of increased prostaglandin synthesis and its subsequent inhibition on the distribution of blood flow intrarenally. Thus, one or more renal prostaglandins, most likely PGE_2, is responsible for mediating increases in blood flow to the renal medulla (24) in response to stimuli as diverse as surgical trauma, hemorrhagic hypotension, and salt loading. A balanced mechanism seems to regulate the distribution of renal blood flow: PGE_2 increases blood flow to the inner cortex and medulla and angiotensin I decreases it. The studies of *Itskovitz and McGiff* (44) have resulted in identifying complementary intrarenal hormones, angiotensin I and PGE_2, which function as local constrictor and dilator hormones, respectively. As a corollary, the nephropathy associated with excessive use of anti-inflammatory-analgesic agents may be initiated by renal medullary ischemia resulting from inhibition of cyclo-oxygenase (45).

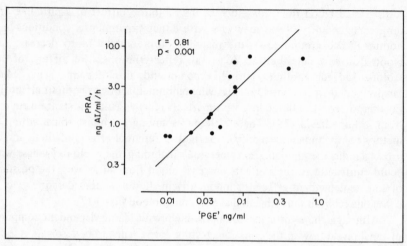

Fig. 3. The relationship between plasma renin activity (PRA) on the ordinates, and on the abscissa, the log of the concentration of 'PGE' in renal venous blood of conscious, anesthetized and surgically stressed dogs. PRA is expressed in nanogram of angiotensin I generated per milliliter of plasma after 1 h of incubation. Concentrations of 'PGE' are expressed in ng/ml blood.

Angiotensin II was the first vasoactive hormone which was demonstrated to release prostaglandins from the kidney (1). Increased release of renin and subsequent generation of angiotensin evoked by acute experimental procedures contribute to increased renal prostaglandin levels. The level of activity of the renin-angiotensin system is highly correlated with renal prostaglandin levels as indicated by corresponding changes in the concentrations of plasma renin activity and 'PGE' in renal venous blood (fig. 3). This correlation does not obtain for 'PGF'. Prostaglandins of the E series can be shown to oppose the renal effects of angiotensin whereas those of the F series cannot (24). Increased production of PGE_2, a potent renal vasodilator (24), helps restore renal blood flow towards its resting level in the face of those stimuli which activate the renin-angiotensin system. Some of those stimuli which release renin may operate through a prostaglandin mechanism (7).

Prostaglandins and Renin Release

Although it had been known since 1970 (1) that angiotensin can release prostaglandins from the kidney, the converse, an effect of the renal prosta-glandin system on renin release, was not appreciated until 1974 when *Larsson et al.* (7) reported that increased prostaglandin synthesis induced by administration of the prostaglandin precursor, arachidonic acid, raised plasma renin activity; whereas, inhibition of prostaglandin synthesis with indomethacin decreased plasma renin activity. This study constitutes the theoretical basis for the administration of aspirin-like drugs in the treatment of Bartter's syndrome (46, 47).

The importance of renal prostaglandins to the functional abnormalities of Bartter's syndrome is suggested by their partial correction (46, 47) after administration of indomethacin, an inhibitor of prostaglandin synthesis (28). Thus, pressor responsiveness to angiotensin and norepinephrine is restored and salt and water loss is reduced. However, as the effects of aspirin-like drugs are not selective, e.g., they also alter the activity of the kallikrein-kinin system (48) and inhibit phosphodiesterase (49), it need not follow that the effects of indomethacin in Bartter's syndrome constitute evidence for a prostaglandin mechanism. Perhaps the singly most important effect of indomethacin in the treatment of these patients is the inhibition of renin release (7). Unraveling the skeins of the complex relationship between the renin-angiotensin system and prostaglandins, particularly as affected by indomethacin, is useful for it may lead not only to therapeutic strategies for the treatment of Bartter's syndrome, but in addition should contribute to an understanding of the mechanisms which regulate renin release and, thereby, contribute to elucidation of the pathophysiology of hypertension and of the physiology of salt and water balance. The most acceptable explanation for the beneficial effects of indomethacin in Bartter's syndrome may be resolved in terms of a single effect of the drug on prostaglandin synthesis (28). The first and most obvious effect of indomethacin is to prevent prostaglandin release in response to angiotensins and kinins within the kidney and other tissues such as the vascular wall. The second effect does not follow from the first but operates because a prostaglandin mechanism participates in the regulation of renin release (7) as well as the release of other renal enzymes such as erythropoietin (50) and, perhaps, kallikrein. This action of indomethacin on renin release may be most important to the treatment of Bartter's syndrome as it will correct the elevated levels of angiotensin and, thereby, prevent the aldosteronism and excessive production of prostaglandins (46). As kallikrein excretion is decreased after administration of indomethacin (51), the possibility that the release of kallikrein is also dependent upon a prostaglandin mechanism should be considered.

Reduction of plasma renin activity in response to indomethacin may be direct, that is via a prostaglandin mechanism, or indirect such as secondary to the effects of the drug on sodium chloride excretion. Changes in sodium chloride composition of the fluid in the distal nephron is an important factor affecting renin release (52). As indomethacin causes retention of salt and water (24), its effect on renin release may be secondary to this action. In order to eliminate a renal functional effect of prostaglandins, *Weber et al.* (8) have used slices of kidney to study the renin-releasing action of prostaglandins. The principal

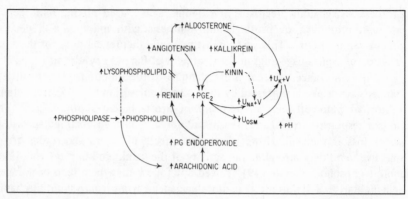

Fig. 4. Relationships among the products of prostaglandin synthesis and renin-angiotensin and kallikrein-kinin systems. These are depicted as they might effect each other and, thereby, urinary excretion of sodium (U_{NA}+V) and potassium (U_K+V), urinary osmolality (U_{OSM}) and plasma pH. The dotted line to lysophospholipid indicates a hypothetical relationship.

product of renal prostaglandin synthesis, PGE_2, which appears in large amounts in the urine of patients with Bartter's syndrome, did not affect renin release directly. However, an intermediate, the cyclic endoperoxide, PGG_2, was shown to increase renin release from renal slices. Prostacyclin and thromboxanes, which also arise from the endoperoxides, were not tested. This study suggests that indomethacin decreases renin release consequent to an effect on a prostaglandin mechanism, probably by elimination of a product of prostaglandin synthesis such as a prostaglandin endoperoxide, or perhaps prostacyclin which has recently been reported to be produced by the kidney (53). If this explanation proves correct, the next question which arises is: why is there increased activity of a prostaglandin mechanism effecting renin release? This leads to a consideration of the rate-limiting step in renal prostaglandin synthesis; namely, arachidonic acid delivery to prostaglandin synthetase (54). Arachidonic acid is released from tissue stores, primarily triglycerides and phospholipids, by acylhydrolases of which phospholipase is probably the most important in most tissues (55). Thus, the functional implications of the findings that prostaglandin-dependent renin release is enhanced points to an antecedent abnormality in the regulation of a tissue lipase which governs substrate delivery to the prostaglandin synthetic machinery.

There is an additional factor related to prostaglandin precursors and their metabolism which has potentially important therapeutic and etiologic implications to Bartter's syndrome; namely the generation of lysophospholipid by-products during prostaglandin synthesis (fig. 4). These substances, which have been reported to inhibit renin (56), are formed when arachidonic acid is released from phospholipids; that is, they are phospholipids minus a fatty acid. Phospholipids are then reconstituted by reincorporation of a fatty acid into the lysophospholipid. An agent which prevents this reaction, the reacylation step, would increase the levels of lysophospholipids. As indomethacin may interfere with reacylation (57), this effect of the drug may be alternative or supplementary to inhibition of prostaglandin synthesis. This analysis leads to the proposal that in Bartter's syndrome there may be an abnormality of the mechanism regulating acylation of the lysophospholipid; this factor could be ancillary or in some cases the major factor which determines hyperreninemia. Despite marked elevation of the activity of the renin-angiotensin system, these patients are not hypertensive presumably because products of cyclo-oxygenase, such as PGE_2, inhibit angiotensins. The proposal that the prostaglandin system contributes to regulation of blood pressure by buffering those stimuli which elevate it has received considerable support (58, 59) although there is one species, the rat, which appears to be an exception.

Prostaglandins: Prohypertensive in the Rat?

There is an important difference in species response observed by *Malik and McGiff* (23) which has raised major questions concerning the suitability of the rat as a model for investigating the role of vasodepressor systems in regulating blood pressure. Prostaglandins of the E series augment the renal vasoconstrictor responses of the isolated perfused kidney of the rat to nerve stimulation in concentrations which do not affect vascular tone (100 pg/ml) and constrict the renal vasculature in higher concentrations. Under identical conditions in the isolated perfused kidney of the rabbit, PGE compounds inhibit adrenergic vasoconstriction and dilate blood vessels. Concentrations of PGE_2 in the perfusing medium as low as 20 pg/ml frequently reduce the vasoconstrictor response to nerve stimulation of the rabbit kidney. At this concentration it is appropriate to invoke the basal state and to consider the physiological relevance of these observations to the regulation of the renal circulation. The anomalous response of the isolated kidney of the rat to PGE_2 takes on greater significance in view of two additional observations. First, *in vivo* PGE_2 can be shown to constrict the renal vasculature of the rat albeit in very high doses. In this regard it should be noted that the renal vascular bed of the rat is peculiarly resistant to PGE_2 when compared to other species as dog and rabbit. However, as noted previously in terms of a proposed role for PGE_2 as a modulator of adrenergic neurotransmitter release, the effects of prostaglandins as modulators may be demonstrated at concentrations 1/250 those which have a direct vascular action. It is this activity as modulators, whereby small amounts of prostaglandins influence the action of hormones for periods in excess of their known biological activity, that compels our attention, particularly in terms of their interactions with vasoactive hormones. The second observation which forces consideration of the significance of the paradoxical vascular response to PGE_2 in the rat isolated kidney is that the principal product of prostaglandin synthetase has the same effect as administered PGE_2. Thus, enhanced synthesis of renal prostaglandins evoked by arachidonic acid administration in the isolated kidney resulted in vasoconstriction and inhibition of prostaglandin synthesis with indomethacin-produced vasodilation (23). It should be recalled that PGE_2 has been identified as the principal product of prostaglandin synthesis in both the rat and rabbit kidneys. Nor are these observations restricted to *in vitro* experimental conditions, as PGE_2 constricts the renal vasculature of the rat *in vivo* (60) as indicated previously.

The study of *Malik and McGiff* (23) urges consideration that renal prostaglandins contribute to the development of hypertension in the rat as an increase in renal vascular resistance by itself might initiate hypertension (61). *Armstrong et al.* (62) have explored this possibility in the New Zealand strain of the genetically hypertensive (GH) rat. Comparison of the vasoconstrictor responses to norepinephrine revealed that the renal vascular bed of GH rats was about 50% more sensitive to norepinephrine than that of normotensive rats and was more susceptible to indomethacin-induced attenuation of the constrictor action of norepinephrine. PGE_2 in doses (100–300 pg/ml) which were without a direct effect on renal blood vessels restored to preindomethacin levels the vasoconstrictor responses to low doses of norepinephrine. Thus, increased renal vascular reactivity to pressor hormones in the GH rat could result partly from elevated levels of prostaglandins. If increased levels of prostaglandin occur, either increased synthesis or decreased degradation of prostaglandin must be responsible. Production of prostaglandins by homogenates of kidneys of GH rats was not different from normal as indicated by measurements of conversion of labeled arachidonic acid to prostaglandins. However, important differences were found between the kidneys of GH and normotensive rats in their abilities to degrade prostaglandins to their 15-keto metabolites, the first and most important step in the metabolism of prostaglandins.

As modulators prostaglandins may amplify or attenuate the vascular actions of hormones or neurotransmitters in concentrations which do not affect blood vessels directly. There is an additional and less well-known corollary to the range of interactions between modulators and hormones, and one which has bearing on the concept that hypertension may develop in the face of normal activity of either the adrenergic nervous system or the renin-angiotensin system. That is, that an abnormality of the modulator, by itself, could result in elevated blood pressure; this may be the case in the GH rat. Thus, the primary or initiating event would be increased levels of PGE_2, or a similar product of prostaglandin synthetase, which results from a deficiency of the major catabolizing enzyme in the face of normal rates of prostaglandin synthesis. It is likely that this disturbance need only occur in the kidney for the development of hypertension in the rat.

Summary

Prostaglandins modulate the effects of vasoactive hormones by attenuating the renal actions of the renin-angiotensin system and contributing to and, perhaps, mediating some of those of the kallikrein-kinin system. A prostaglandin mechanism participates in the regulation of renin and erythropoietin release. When renal function is compromised, the circulation to the kidney is sustained by a major prostaglandin component withdrawal of which results in significant hemodynamic effects, particularly reduction of blood flow to the inner cortex and medulla.

Acknowledgments

We thank Ms. *J. Lariviere* for typing the manuscript and Dr. *P. Baer* for his helpful suggestions. This work was supported in part by USPHS Grant HL-18845.

References

1 *McGiff, J.C.; Crowshaw, K.; Terragno, N.A., and Lonigro, A.J.:* Release of a prostaglandin-like substance into renal venous blood in response to angiotensin II. Circulation Res. *26/27:* suppl. I, pp. 121–130 (1970).

2 *McGiff, J.C.; Crowshaw, K.; Terragno, N.A., and Lonigro, A.J.:* Renal prostaglandins: possible regulators of the renal actions of pressor hormones. Nature, Lond. *227:* 1255–1257 (1970).

3 *McGiff, J.C.; Crowshaw, K.; Terragno, N.A.; Malik, K.U., and Lonigro, A.J.:* Differential effect of noradrenaline and renal nerve stimulation on vascular resistance in the dog kidney and the release of a prostaglandin E-like substance. Clin. Sci. *42:* 223–233 (1972).

4 *Terragno, D.A.; Crowshaw, K.; Terragno, N.A., and McGiff, J.C.:* Prostaglandin synthesis by bovine mesenteric arteries and veins. Circulation Res. *36/37:* suppl. I, pp. 76–80 (1975).

5 *Needleman, P.; Kulkarni, P.S., and Raz, A.:* Coronary tone modulation: formation and actions of prostaglandins, endoperoxides, and thromboxanes. Science *195:* 409–412 (1977).

6 *Moncada, S.; Gryglewski, R.; Bunting, S., and Vane, J.R.:* An enzyme isolated from arteries transforms prostaglandin endoperoxides to an unstable substance that inhibits platelet aggregation. Nature, Lond. *263:* 663–665 (1976).

7 *Larsson, C.; Weber, P.; Änggard, E.:* Arachidonic acid increases and indomethacin decreases plasma renin activity in the rabbit. Eur. J. pharmacol. *28:* 391–394 (1974).

8 Weber, P.; Larsson, C.; Änggard, E.; Hamberg, M.; Corey, E.J.; Nicolaou, K.C., and Samuelsson, B.: Stimulation of renin release from rabbit renal cortex by arachidonic acid and prostaglandin endoperoxides. Circulation Res. 39: 868–874 (1976).

9 Page, I.H. and McCubbin, J.W.: Renal hypertension, p. 56 (Year Book, Chicago 1968).

10 Crowshaw, K. and Szlyk, J.Z.: Distribution of prostaglandins in rabbit kidney. Biochem. J. 116: 421–424 (1970).

11 Larsson, C. and Änggard, E.: Regional differences in the formation and metabolism of prostaglandins in the rabbit kidney. Eur. J. Pharmacol. 21: 30–36 (1973).

12 Muirhead, E.E.; Germain, G.; Leach, B.E.; Pitcock, J.A.; Stephenson, P.; Brooks, B.; Brosius, W.L.; Daniels, E.G., and Hinman, J.W.: Production of renomedullary prostaglandins by renomedullary interstitial cells grown in tissue culture. Circulation Res. 31: suppl. II, pp. 161–172 (1972).

13 Kauker, M.L.: Prostaglandin E$_2$ effect from the luminal side on renal tubular ^{22}Na efflux: tracer microinjection studies (39653). Proc. Soc. exp. Biol. Med. 154: 274–277 (1977).

14 Itskovitz, H.D.; Terragno, N.A., and McGiff, J.C.: Effect of a renal prostaglandin on distribution of blood flow in the isolated canine kidney. Circulation Res. 34: 770–776 (1974).

15 Kauker, M.L.: Tracer microinjection studies of prostaglandin E$_2$ transport in the rat nephron. J. Pharmac. exp. Ther. 193: 274–280 (1975).

16 Janszen, F.H.A. and Nugteren, D.H.: Histochemical localisation of prostaglandin synthetase. Histochemie 27: 159–164 (1971).

17 McGiff, J.C.; Itskovitz, H.D., and Terragno, N.A.: The actions of bradykinin and eledoisin in the canine isolated kidney: relationships to prostaglandins. Clin. Sci. molec. Med. 49: 125–131 (1975).

18 Grantham, J.J. and Orloff, J.: Effect of prostaglandin E$_1$ on the permeability response of the isolated collecting tubule to vasopressin, adenosine 3',5'-monophosphate, and theophylline. J. clin. Invest. 47: 1154–1161 (1968).

19 Carretero, O.A. and Scicli, A.G.: Renal kallikrein: its localization and possible role in renal function. Fed. Proc. Fed. Am. Socs exp. Biol. 35: 194–198 (1976).

20 McGiff, J.C.; Terragno, N.A.; Malik, K.U., and Lonigro, A.J.: Release of a prostaglandin E-like substance from canine kidney by bradykinin. Circulation Res. 31: 36–43 (1972).

21 Gryglewski, R.J.; Bunting, S.; Moncada, S.; Flower, R.J., and Vane, J.R.: Arterial walls are protected against deposition of platelet thrombi by a substance (prostaglandin X) which they make from prostaglandin endoperoxides. Prostaglandins 12: 685–713 (1976).

22 Danon, A.; Chang, L.C.T.; Sweetman, B.J.; Nies, A.S., and Oates, J.A.: Synthesis of prostaglandins by the rat renal papilla in vitro. Mechanism of stimulation by angiotensin II. Biochim. biophys. Acta 388: 71–83 (1975).

23 Malik, K.U. and McGiff, J.C.: Modulation by prostaglandins of adrenergic transmission in the isolated perfused rabbit and rat kidney. Circulation Res. 36: 599–609 (1975).

24 McGiff, J.C. and Itskovitz, H.D.: Prostaglandins and the kidney. Circulation Res. 33: 479–488 (1973).

25 Hamberg, M.; Svensson, J., and Samuelsson, B.: Thromboxanes: a new group of biologically active compounds derived from prostaglandin endoperoxides. Proc. natn. Acad. Sci. USA 72: 2994–2998 (1975).

26 Morrison, A.; Nishikawa, K., and Needleman, P.: Unmasking of thromboxane A$_2$ synthesis by ureteral obstruction in the rabbit kidney. Nature, Lond. 267: 259–260 (1977).

27 Yarger, W.E. and Griffith, L.D.: Intrarenal hemodynamics following chronic unilateral ureteral obstruction in the dog. Am. J. Physiol. 227: 816–826 (1974).

28 Vane, J.R.: Inhibition of prostaglandin synthesis as a mechanism of action for aspirin-like drugs. Nature, Lond. 231: 232–235 (1971).

29 Needleman, P.; Moncada, S.; Raz, A.; Ferrendelli, J.A., and Minkes, M.: Application of imidazole, as a selective thromboxane synthetase inhibitor in human platelets. Proc. natn. Acad. Sci. USA 74: 1716–1720 (1977).

30 Moncada, S.; Gryglewski, R.J.; Bunting, S., and Vane, J.R.: A lipid peroxide inhibits the enzyme in blood vessel microsomes that generates from prostaglandin endoperoxides the substance (prostaglandin X) which prevents platelet aggregation. Prostaglandins 12: 715–737 (1976).

31 Terragno, N.A.; Terragno, D.A., and McGiff, J.C.: Contribution of prostaglandins to the renal circulation in conscious, anesthetized and laparotomized dogs. Circulation Res. (in press, 1977).

32 Zins, G.R.: Renal prostaglandins. Am. J. Med. 58: 14–24 (1975).

33 Swain, J.A.; Heyndrickx, G.R.; Boettcher, D.H., and Vatner, S.F.: Prostaglandin control of renal circulation in the unanesthetized dog and baboon. Am. J. Physiol. 229: 826–830 (1975).

34 Terragno, N.A.; Terragno, D.A.; Pacholczyk, D., and McGiff, J.C.: Prostaglandins and the regulation of uterine blood flow in pregnancy. Nature, Lond. 249: 57–58 (1974).

35 Data, J.L.; Chang, L.C.T., and Nies, A.S.: Alteration of canine renal vascular response to hemorrhage by inhibitors of prostaglandin synthesis. Am. J. Physiol. 230: 940–945 (1976).

36 Pace-Asciak, C. and Cole, S.: Inhibitors of prostaglandin catabolism. I. Differential sensitivity of 9-PGDH, 13-PGR and 15-PGDH to low concentrations of indomethacin. Experientia 31: 143–145 (1975).

37 Beilin, L.J. and Bhattacharya, J.: The effects of prostaglandin synthesis inhibitors on renal blood flow distribution within the kidney. J. Physiol. 256: 9P–10P (1976).

38 Vatner, S.F.: Effects of hemorrhage on regional blood flow distribution in dogs and primates. J. clin. Invest. 54: 225–235 (1974).

39 Lonigro, A.J.; Itskovitz, H.D.; Crowshaw, K., and McGiff, J.C.: Dependency of renal blood flow on prostaglandin synthesis in the dog. Circulation Res. 32: 712–717 (1973).

40 Itskovitz, H.D.; Stemper, J.; Pacholcyzk, D., and McGiff, J.C.: Renal prostaglandins: determinants of intrarenal distribution of blood flow. Clin. Sci. molec. Med. 45: suppl. I, pp. 321–324 (1973).

41 Larsson, C. and Änggard, E.: Increased juxtamedullary blood flow on stimulation of intrarenal prostaglandin biosynthesis. Eur. J. Pharmacol. 25: 326–334 (1974).

42 Kirschenbaum, M.A.; White, N.; Stein, J.H., and Ferris, T.F.: Redistribution of renal cortical blood flow during inhibition of prostaglandin synthesis. Am. J. Physiol. 227: 801–805 (1974).

43 Brown, J.J.; Davies, D.L.; Lever, A.F.; Parker, R.A., and Robertson, J.I.S.: Assay of renin in single glomeruli: renin distribution in the normal rabbit kidney. Lancet ii: 668 (1963).

44 Itskovitz, H.D. and McGiff, J.C.: Hormonal regulation of the renal circulation. Circulation Res. 34/35: suppl. I, pp. 65–73 (1974).

45 Nanra, R.S.; Chirawong, P., and Kincaid-Smith, P.: Medullary ischemia in experimental analgesic nephropathy. The pathogenesis of renal papillary necrosis. Aust. N. Z. Jl Med. 3: 580–586 (1973).

46 Gill, J.R.; Frölich, J.C.; Bowden, R.E.; Taylor, A.A.; Keiser, H.R.; Seyberth, H.W.;

Oates, J.A., and Bartter, F.C.: Bartter's syndrome: a disorder characterized by high urinary prostaglandins and a dependence of hyperreninemia on prostaglandin synthesis. Am. J. Med. *61:* 43–51 (1976).

47 *Verberckmoes, R.; Damme, B. van; Clement, J.; Amery, A., and Michielsen, P.:* Bartter's syndrome with hyperplasia of renomedullary cells: successful treatment with indomethacin. Kidney int. *9:* 302–307 (1976).

48 *Lewis, G.P.:* Kinins in inflammation and tissue injury; in *Erdos* Handbook of experimental pharmacology, vol. 25: Bradykinin, kallidin and kallikrein, p. 525 (Springer, Berlin 1970).

49 *Flores, A.G.A. and Sharp, G.W.G.:* Endogenous prostaglandins and osmotic water flow in the toad bladder. Am. J. Physiol. *233:* 1392–1397 (1972).

50 *Mujovic, V.M. and Fisher, J.W.:* The effects of indomethacin on erythropoietin production in dogs following renal artery constriction. I. The possible role of prostaglandins in the generation of erythropoietin by the kidney. J. Pharmac. exp. Ther. *191:* 575–580 (1974).

51 *Vinci, J.M.; Telles, D.A.; Bowden, R.E.; Izzo, J.L.; Keiser, H.R.; Radfar, N.; Taylor, A.A.; Gill, J.R., and Bartter, F.C.:* The kallikrein-kinin system (KKS) in Bartter's syndrome (BS) and its response to prostaglandin synthetase inhibition (PGSI). Abstract. Clin. Res. *24:* 414A (1976).

52 *Thurau, K.; Schnermann, J.; Nagel, W.; Horster, M., and Wahl, M.:* Composition of tubular fluid in the macula densa segment as a factor regulating the function of the juxtaglomerular apparatus. Circulation Res. *20/21:* suppl. II, pp. 79–90 (1967).

53 *Frölich, J.C.:* Public communication on novel products of prostaglandin synthesis. Winter Prostaglandin Conf., Vail, Colo. 1977.

54 *Hinman, J.W.:* Prostaglandins. A. Rev. Biochem. *41:* 161–178 (1972).

55 *Hong, S.L. and Levine, L.:* Stimulation of prostaglandin synthesis by bradykinin and thrombin and their mechanisms of action on MC5-5 fibroblasts. J. biol. Chem. *251:* 5814–5816 (1976).

56 *Overturf, M.; Druilhet, R.E., and Kirkendall, W.M.:* Effect of human kidney lipids on human kidney renin activity. Biochem. Pharmac. *25:* 2443–2453 (1976).

57 *Flower, R.J. and Blackwell, G.J.:* The importance of phospholipase-A_2 in prostaglandin biosynthesis. Biochem. Pharmac. *25:* 285–291 (1976).

58 *Vane, J.R. and McGiff, J.C.:* Possible contributions of endogenous prostaglandins to the control of blood pressure. Circulation Res. *36/37:* suppl. I, pp. 68–75 (1975).

59 *Colina-Chourio, J.; McGiff, J.C., and Nasjletti, A.:* Development of high blood pressure following inhibition of prostaglandin synthesis. Abstract. Fed. Proc. Fed. Am. Socs exp. Biol. *34:* 368 (1975).

60 *Baer, P.:* Personal commun.

61 *Tobian, L.:* How sodium and the kidney relate to the hypertensive arteriole. Fed. Proc. Fed. Am. Socs exp. Biol. *33:* 138–142 (1974).

62 *Armstrong, J.M.; Blackwell, G.J.; Flower, R.J.; McGiff, J.C.; Mullane, K.M., and Vane, J.R.:* Genetic hypertension in rats is accompanied by a defect in renal prostaglandin catabolism. Nature, Lond. *260:* 582–586 (1976).

J.C. McGiff, Department of Pharmacology, University of Tennessee, Center for the Health Sciences, 800 Madison Avenue, *Memphis, Tenn.* (USA)

Contr. Nephrol., vol. 12, pp. 41–53 (Karger, Basel 1978)

Antihypertensive Effect of Volume Depletion: Interrelation with Renal Prostaglandins

R. Düsing, A. Attallah, W.E. Braselton and J.B. Lee

Department of Medicine, E.J. Meyer Memorial Hospital, State University of New York at Buffalo, Buffalo, N.Y. and the Medical College of Georgia, Augusta, Ga.

The discovery, in 1962, of biological activity ultimately attributable to the renal prostaglandins was the result of a search for antihypertensive principles within the kidney consonant with a possible antihypertensive endocrine function of this organ (15). Although an exhaustive literature has appeared outlining the possible prohypertensive roles of the kidney, most notably the sympathetic nervous system and the renal renin-angiotensin system, to date there is no solid evidence to implicate a primary overactivity of either prohypertensive function in the etiology of human essential hypertension. Since the early investigations of *Muirhead et al.* (20) suggested that the renal medulla exerted an antihypertensive effect in the prevention of acute canine salt-dependent renoprival hypertension, extracts of rabbit renal medulla were injected into the anesthetized rat (15, 16) to see if any blood pressure lowering effect would be observed. Figure 1 illustrates that rabbit renal medullary extracts produce an immediate fall in blood pressure with a subsequent return to control levels over a period of 10–15 min. Ultimately from 20 kg of rabbit renal medulla, three compounds were identified, two of which were responsible for the blood pressure lowering effect in the anesthetized rat (13).

These compounds, originally called compound 1, compound 2 and medullin, were identified by mass-spectrometry as $PGF_{2\alpha}$, PGE_2 and a new prostaglandin class, the PGA_2s (fig. 2) (14). It soon became evident that the PGE and PGA compounds not only lowered blood pressure but also had notable effects in enhancing renal blood flow and sodium and water excretion (8, 10, 24). Subsequently, a voluminous literature has appeared, often controversial in nature, in attempts to delineate possible physiological roles for the renal prostaglandins in blood pressure regulation, regulation of renal blood flow and salt and water homeostasis. Since these functions are closely interrelated and interdependent, investigations of a possible role of prostaglandins in concert with the host of other humoral and vascular factors (kinins, vasopressin, norepinephrine, etc.) known to affect such functions have proven to be a most formidable task.

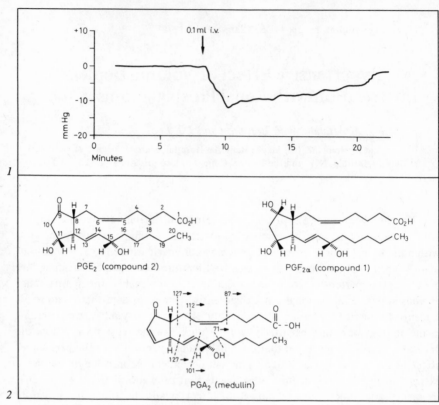

Fig. 1. Sustained vasodepressor effect in the pentolium-treated rat following incubation with slices of rabbit renal medulla. From *Lee et al.* (16) with permission of the publisher.

Fig. 2. Identification of compound 1 as $PGF_{2\alpha}$, compound 2 as PGE_2 and medullin as PGA_2, by mass-spectroscopy. Suggested fragmentation for all three compounds are indicated in the structure of PGA_2. From *Lee et al.* (14) with permission of the publisher.

With regard to the heterogeneity of prostaglandin functions, this discussion will be limited to only one aspect of this multifactorial problem, namely, a consideration of the mechanisms whereby manipulations which tend to produce volume depletion result in a lowering of blood pressure. Since volume depletion produces a marked activation of the prohypertensive function of the kidney, notably renin-angiotensin-aldosterone system, it might be expected that the blood pressure would rise rather than fall in this situation. To date efforts to explain this paradox have centered around two lines of evidence: first, a relative insensitivity to the pressor effects of either norepinephrine or angiotensin in volume depletion and, secondly, intracellular shifts of sodium and potassium at the vascular site of relaxation in this condition. This discussion is limited to a

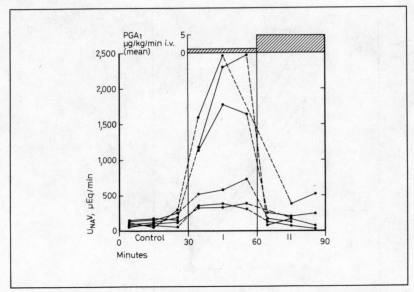

Fig. 3. Effect of PGA₁ administered intravenously in 6 of 20 patients with essential hypertension. Each period (control, I and II) consisted of three 10-min clearance observations. Note increase in sodium excretion in period I (before any change in blood pressure) and subsequent decline to control when normotension was achieved in period II. From *Lee et al.* (17) with permission of the publisher.

brief consideration of the possibility that PGA or the PGE compounds may participate in the mechanism whereby blood pressure falls during the state of volume depletion. Figure 3 illustrates that when PGA₁ is infused into patients with essential hypertension at a rate of approximately 1 μg/kg/min it is evident that there is a marked, albeit transient, natriuresis (period I) which is accompanied by an increase in renal blood flow (not shown). During period I the blood pressure in all subjects remained elevated. However, when the PGA₁ infusion rate was increased to 5 μg/kg/min in period II, the blood pressure fell to normotensive levels, the result of peripheral arteriolar dilation. When this occurred, it is evident from figure 3 that sodium excretion (and renal blood flow) returned to control values. In this sense, the PGA compounds act as 'ideal' antihypertensive agents in that they favorably affect renal resistance, sodium and water homeostasis, total peripheral resistance, and ultimately blood pressure (17, 25). Similar results have been described independently for PGA₁ by *Carr* (1) and for PGA₂ by *Hornych et al.* (7). From these observations and observations in animals, we proposed that perhaps the PGA and PGE compounds were natriuretic factors, and in fact, might at least in part be responsible for the activity of the so-called natriuretic factor (11). The first evidence that this was not true in

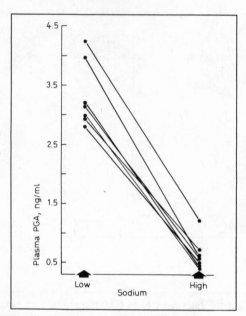

Fig. 4. Individual changes in plasma PGA in response to low and high sodium intakes. Low salt diet: 50 mEq/24 h, high salt diet: greater than 175 mEq/24 h. For low versus high sodium intake p <0.001. From *Payakkapan et al.* (23) with permission of the publisher.

normotensive humans (especially for the PGA compounds) was presented independently by *Zusman et al.* (26, 27) and ourselves (12, 23) showing that under conditions of a low sodium intake circulating immunoreactive PGA in the plasma rises significantly.

In 65 healthy medical students and nurses with sodium intakes varying between 20 and 180 mEq/24 h immunoreactive PGA rose markedly under conditions of the low sodium intake. Figure 4 illustrates that in the same individual PGA rose significantly during a low sodium intake of 50 mEq/24 h when compared with values during a high sodium intake of 170 mEq/24 h. These results for all practical purposes excluded a circulating natriuretic role for immunoreactive PGA although an intrarenal natriuretic role for the PGE compounds still remains a very distinct possibility.

In the same 65 normotensive human subjects, plasma renin activity and urinary aldosterone rose significantly under conditions of low sodium intake, in concert with the rise in immunoreactive PGA (23). The parallel rise in the antihypertensive prostaglandins and the prohypertensive renin-angiotensin system were difficult to reconcile until the studies of *Golub et al.* (4) who showed that infusion of PGA_1 in normotensive human subjects produced a significant rise in plasma renin activity and plasma aldosterone.

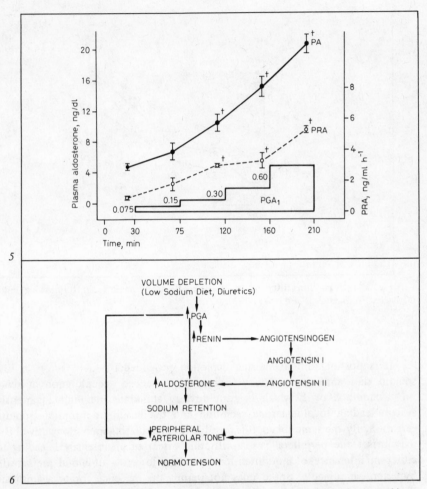

Fig. 5. The effect of PGA$_1$ on renin and aldosterone in normotensive human subjects. From *Golub et al.* (4) with permission of the publisher.

Fig. 6. Hypothetical schema of the possible interaction of increased plasma PGA in the biosynthesis of renin-angiotensin-aldosterone production and maintenance of normotension in normotensive subjects. From *Payakkapan et al.* (23) with permission of the publisher.

It is evident from figure 5 that at infusion rates between 0.075 and 0.60 μg/ kg/min, conditions where no change in blood pressure or volume depletion occur, plasma renin activity rose in a highly significant fashion and that this rise was accompanied by a significant elevation in plasma aldosterone. These studies have led to the formulation of a hypothesis of the role of prostaglandins in the antihypertensive effect of volume depletion as shown in figure 6.

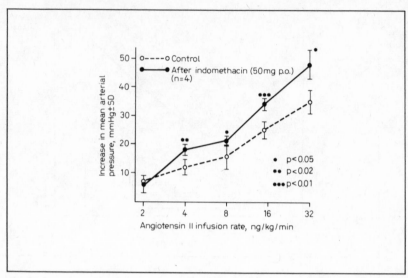

Fig. 7. Increase sensitivity to the pressor effects of angiotensin II following indomethacin. From data of *Guthrie et al.* (5).

It is postulated that volume depletion produced by such factors as low sodium diet and diuretics may lead to an enhanced accumulation of renal prostaglandin A or E which in turn directly stimulate the renin-angiotensin system leading to aldosterone secretion and the ensuing appropriate sodium retention. By the same token, it is possible that circulating immunoreactive PGA may offset the peripheral vasoconstricting effect of angiotensin II leading to either maintenance of normotension or actual lowering of blood pressure in hypertensive subjects. Space does not permit the evidence for or against the actual operation of this hypothesis, with the exception that one of the previous explanations for the beneficial effects of volume depletion on blood pressure, that is the relative insensitivity to norepinephrine or angiotensin, may be explained by the fact that the administration of the prostaglandin synthetase inhibitor indomethacin results in a return of decreased sensitivity to the pressor effects of these agents to values observed under conditions of volume expansion. *Guthrie et al.* (5) have shown that in normal human subjects the administration of indomethacin leads to an enhanced sensitivity to the blood pressure elevating effects of various infusion rates of angiotensin II and norepinephrine (fig. 7). These findings support the hypothesis that endogenous prostaglandin synthesis may be an important determinant of vascular reactivity to pressor agents in humans.

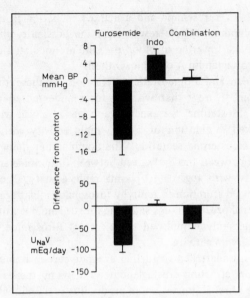

Fig. 8. Effects of furosemide, indomethacin and a combination of these agents on mean arterial blood pressure and daily sodium excretion. Bars denote mean increase or decrease as compared with that in the control period of pooled data on 4 normal subjects and 6 hypertensive patients with 1 SEM. The results were statistically significant except for (1) the effect of combination treatment on mean blood pressure and (2) the effect of indomethacin treatment on daily urinary sodium excretion. From *Patak et al.* (21) with permission of the publisher.

In an effort to determine whether the ameliorative effects of a low sodium diet on blood pressure, possibly mediated by prostaglandins, could also be extended to other volume-depleting stimuli such as diuretic therapy, *Patak et al.* (21) studied 15 normotensive subjects and patients with essential hypertension for 30 days on a 150 mEq sodium diet. Following a 4-day control period furosemide was administered at a dose of 80 mg every 8 h for 4 days. Following a 4-day equilibrium period, indomethacin was then administered 50 mg orally every 6 h for an additional 4 days. Again, following a second 4-day equilibration period a combination of furosemide and indomethacin in the same dosage was administered.

Figure 8 shows that in all patients there was a significant reduction of mean blood pressure of approximately 15 mm Hg with furosemide and that this fall in blood pressure was accompanied by a significant natriuresis. Administration of indomethacin resulted in a slight but highly significant rise in blood pressure in all subjects studied and was without effect on urinary sodium excretion. The combination of furosemide and indomethacin resulted in an abolition of all the blood pressure lowering effects of furosemide and a marked reduction in the natriuretic activity of this compound. These results raise the possibility that furosemide-induced natriuresis and lowering of blood pressure are mediated by enhanced renal or extrarenal prostaglandin A or E biosynthesis.

In addition to the vascular and homeostatic interaction of these two compounds, it is evident from figure 9 that whereas furosemide led to an elevation in renin and urinary aldosterone, the exact reverse was observed with indomethacin, that is, a marked diminution of plasma renin activity with a resultant decrease in urinary aldosterone secretion. The combination therapy resulted in intermediate values between those observed between furosemide and indomethacin alone particularly with regards to plasma renin activity. These results suggested that the enhanced secretion of renin by furosemide might again be mediated by renal or extrarenal prostaglandins since the prostaglandin synthetase inhibitor indomethacin markedly diminished the effect of furosemide in promotion of this rise in plasma renin activity.

Studies therefore have been undertaken to further evaluate certain homeostatic and pharmacologic factors affecting prostaglandin synthesis by the renal medulla which is the major site of renal prostaglandin production (9). Following exsanguination and removal of the kidneys, slices of rabbit renal medulla were incubated in Krebs-Ringer bicarbonate buffer with purified $1\text{-}^{14}\text{C}$-arachidonic acid at concentrations between 1 and 100 μM. Following a 30-min incubation the media were acidified, extracted with ethyl acetate and silicic acid chromatography performed by a refinement allowing separation of arachidonic acid from prostaglandins. Following the elution of the various prostaglandins, the PGA, PGE and PGF peaks were counted and the results expressed as nanomoles of arachidonic acid incorporated into each of the various renal prostaglandins per gram initial wet weight, by methods previously described (18). In some instances radioimmunoassay of total PGE_2 appearing in the media following incubation was performed utilizing the antibody of *Dray and Charbonnel* (3).

The column separation of arachidonic acid from PGA_2, PGE_2 and $PGF_{2\alpha}$ is shown in figure 10. The structures of the prostaglandins in the various peaks were confirmed by GC-mass spectrometry. Although the PGA_2 peak in this chromatograph appears quite distinct from the arachidonic acid peak, it was soon found that following the change of solvents, additional arachidonic acid was also eluted with the PGA_2 and thus further purification of this compound is necessary to quantitate its conversion from arachidonic acid. The results presented in this investigation therefore, will be for PGE_2 and $PGE_{2\alpha}$. Preliminary observations have shown however, that there is a significant biosynthesis of PGA_2 from arachidonic acid which cannot be accounted for by the 2–3% conversion of PGE_2 to PGA_2 during the incubation and subsequent extraction procedure. Thus, although biosynthesis of PGA_2 has not been noted in homogenates of renal papilla utilizing glutathione in the media (2), it appears that

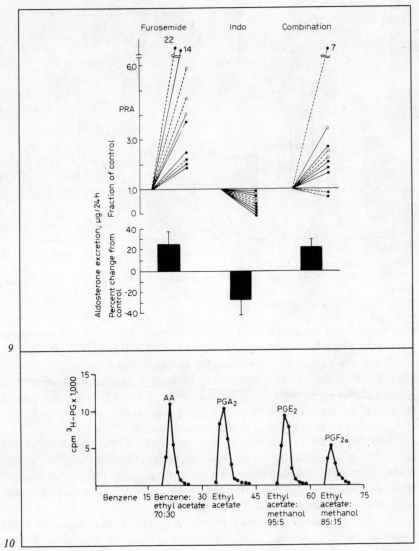

Fig. 9. Effects of furosemide, indomethacin and the combination of these agents on plasma renin activity and urinary aldosterone. Each bar indicates the mean value of a 4-day clearance period. Changes in cumulative sodium excretion produced by each treatment regimen are expressed as mean sodium retention (bar graphs upward) or mean sodium loss (bar graphs downward) per day for each 4-day treatment period compared with the initial control period. Continuous and interrupted lines represent hypertensive and normotensive subjects, respectively. From *Patak et al.* (21) with permission of the publisher.

Fig. 10. Silicic acid chromatography separation of arachidonic acid from PGA_2, PGE_2 and $PGF_{2\alpha}$. Silicic acid 2 g: column 1 $\times$ 4 cm.

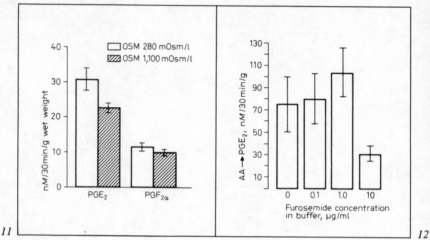

Fig. 11. Effect of sodium chloride-induced osmolality change on the incorporation of arachidonic acid into PGE_2 and $PGF_{2\alpha}$.

Fig. 12. Effect of furosemide on the incorporation of arachidonic acid into PGE_2 and $PGF_{2\alpha}$.

such biosynthesis can readily take place utilizing the intact slice system without glutathione which forms water-soluble sulfhydryl adducts with PGA_2 (6).

There was a linear incorporation of arachidonic acid into PGE_2 between 1 and 200 μM of 0.26 ± 0.05 and 57.0 ± 5.9 μmol/g, respectively. Comparable values for $PGF_{2\alpha}$ were 0.13 ± 0.01 and 20.1 ± 4.9 μmol/g; thus the rate of incorporation of arachidonic acid into PGE_2 was approximately twice that of $PGF_{2\alpha}$ at any given arachidonic acid concentration.

The conversion of arachidonic acid into PGE_2 was conditioned by the concentration of sodium in the medium. It is evident from figure 11 that at high medium osmolality the conversion of arachidonic acid into PGE_2 is significantly diminished while that of $PGF_{2\alpha}$ is unchanged. The significance of this effect of sodium chloride induced osmolality change remains unclear.

In the next series of studies the direct effect of furosemide on PGE_2 synthesis was investigated. It is evident from figure 12 that at a medium concentration of 10 μg/ml, which approximates pharmacologic concentration of furosemide in human plasma and urine, there is no stimulation of the conversion of arachidonic acid into PGE_2. Rather there is a rather marked inhibition of *in vitro* prostaglandin biosynthesis from arachidonic acid by furosemide. When immunoreactive PGE_2 is measured rather than the conversion of radioactive arachidonic acid into PGE_2, it is also evident that there is a significant inhibition of the total production of PGE_2 by slices of rabbit renal medulla under the influence of furosemide *in vitro* (fig. 13).

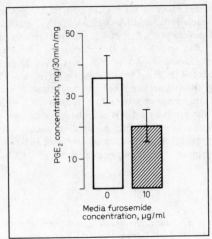

Fig. 13. Effect of furosemide on total immunoreactive PGE_2 production *in vitro.*

In conclusion, although it appears from *in vivo* studies that volume depletion induced by low sodium and furosemide may be mediated by enhanced prostaglandin levels, *in vitro* studies appear to suggest that no direct stimulation of PG synthesis by furosemide occurs. Although accumulation of prostaglandins seem to participate in the *in vivo* effects of furosemide-induced volume depletion, it is possible that these effects are mediated by inhibition of prostaglandin dehydrogenase by this compound as shown by *Paulsrud and Miller* (22). Another possibility is that renal functional changes induced by furosemide lead to a significant rise in renin-angiotensin, which in turn has been shown by *McGiff et al.* (19) to result in an elaboration of PGE_2 by the kidney. A third possibility is that part of the decreased PGE_2 synthesis by furosemide may reflect an accelerated conversion to PGA_2. Verification of the mechanism of the effects of volume depletion whether by sodium deprivation or furosemide involving prostaglandins must await further detailed investigation.

Summary

Since the original studies of *Patak et al.* in 1975 revealed that the antihypertensive and natriuretic effects of furosemide were markedly blunted or abrogated by indomethacin in both normotensive and hypertensive man, it has been postulated that the ameliorative effects of furosemide in human essential hypertension might be mediated by release of intrarenal prostaglandins. To study the direct effects of furosemide on prostaglandin biosynthesis and release, slices of rabbit renal medulla were incubated in Krebs-Ringer bicarbonate buffer, glucose 10 mM, 1-^{14}C-arachidonic acid (AA) 10 μM, HSA 4 g/100 ml, 30 min 37 °C. Measurements were made of radioactive AA → PGE$_2$, and total endogenous immunoreactive PGE$_2$ production (iPGE$_2$) with and without the addition of furosemide (10 μg/ml) to the media. In the absence of furosemide AA → PGE$_2$ was 73 ± 22 nmol/ 30 min/g and in the presence of furosemide it fell to 30 ± 4 nmol/30 min/g. iPGE$_2$ was 33 ± 4 ng/30 min/mg and decreased to 25 ± 3 mg with furosemide. These results indicate that the natriuresis and antihypertensive effect of furosemide *in vivo*, which is associated with a significant increase in urinary PGE$_2$, is not the result of a direct stimulation of furosemide on prostaglandin synthesis but may result from a decrease in PGE metabolism, conversion to another biologically active prostaglandin or possibly be a reflection of events secondary to a direct effect of furosemide on renal hemodynamics and electrolyte excretion.

Acknowledgement

This study was supported in part by a General Research Grant from the State University of New York at Buffalo, Grant No. PN.50967 from the New York State Health Research Council and Astra Pharmaceutical Company. The authors are also indebted to *Geraldine McEwen* and *Ismael Allende* for technical assistance and *Mrs. Arlene Mathews* for secretarial assistance.

References

1 *Carr, A.A.:* Hemodynamic and renal effects of a prostaglandin, PGA$_1$ in subjects with essential hypertension. Am. J. med. Sci. *259:* 21–26 (1970).

2 *Crowshaw, K.:* The incorporation of 1-^{14}C-arachidonic acid into the lipids of rabbit renal slices and conversion to prostaglandins E$_2$ and F$_{2\alpha}$. Prostaglandins *3:* 607–620 (1973).

3 *Dray, F. and Charbonnel, B.:* Radioimmunoassay of prostaglandins Fα and E$_1$ in human plasma; dans Prostaglandines 1973, pp. 133–158 (French) (Inserm, Paris 1973).

4 *Golub, M.S.; Speckart, P.F.; Zia, P.K., and Horton, R.:* The effect of prostaglandin A$_1$ on renin and aldosterone in man. Circulation Res. *39:* 574–579 (1976).

5 *Guthrie, G.P.; Messerli, F.H.; Kuchel, O., and Genest, J.:* Enhanced sensitivity to pressor agents by indomethacin in normal man. Clin. Res. *24:* 220 (1976).

6 *Ham, E.A. and Kuehl, F.A.:* The reaction of PGA$_1$ with sulfhydryl groups: a component in the binding of A-type prostaglandins to proteins. Prostaglandins *10:* 217–229 (1975).

7 *Hornych, A.; Safar, M.; Papanicolaou, N.; Meyer, P., and Milliez, P.:* Renal and cardiovascular effects of prostaglandin A$_2$ in hypertensive patients. Eur. J. clin. Invest. *3:* 391–398 (1973).

8 *Johnston, H.H.; Herzog, J.P., and Lauler, D.P.:* Effect of prostaglandin E$_1$ on renal hemodynamics, sodium and water excretion. Am. J. Physiol. *231:* 939–945 (1967).

9 *Larsson, C. and Änggård, E.:* Regional differences in the formation and metabolism of prostaglandins in rabbit kidney. Eur. J. Pharmacol. *21:* 30–36 (1973).

10 *Lee, J.B.:* Chemical and physiological properties of renal prostaglandins. The antihypertensive effect of medullin in essential hypertension; in Prostaglandins. IInd Nobel Symp., pp. 197–210 (Almqvist & Wiksell, Stockholm 1967).

11 *Lee, J.B.:* Natriuretic 'hormone' and the renal prostaglandins. Prostaglandins *1:* 55–70 (1972).

12 *Lee, J.B.:* Hypertension, natriuresis and the renomedullary prostaglandins: an overview. Prostaglandins *3:* 551–579 (1973).

13 *Lee, J.B.; Covino, B.G.; Rakman, B.H., and Smith, E.R.:* Renomedullary vasodepressor substance. Medullin: isolation, chemical characterization and physiological properties. Circulation Res. *7:* 57–77 (1965).

14 *Lee, J.B.; Crowshaw, K.; Takman, B.H.; Attrep, K.A., and Gougoutas, J.Z.:* Identification of prostaglandins E_2, $F_{2\alpha}$ and A_2 from rabbit kidney medulla. Biochem. J. *105:* 1251–1260 (1967).

15 *Lee, J.B.; Hickler, R.B.; Saravis, C.A., and Thorn, G.W.:* Sustained depressor effects of renomedullary extracts. Circulation *26:* 747 (1962).

16 *Lee, J.B.; Hickler, R.B.; Saravis, C.A., and Thorn, G.W.:* Sustained depressor effects of renal medullary extract in the normotensive rat. Circulation Res. *13:* 359–366 (1963).

17 *Lee, J.B.; McGiff, J.C.; Kannegiesser, H.; Aykent, Y.Y.; Mudd, J.G., and Frawley, T.F.:* Antihypertensive renal effects of prostaglandin A_1 in patients with essential hypertension. Ann. intern. Med. *74:* 703–710 (1971).

18 *Lee, J.B.; Vance, V.K., and Cahill, G.F., jr.:* Metabolism of C^{14}-labeled substrate by rabbit kidney cortex and medulla. Am. J. Physiol. *203:* 27–36 (1962).

19 *McGiff, J.C.; Crowshaw, K.; Terragno, N.A., and Lonigro, A.J.:* Release of a prostaglandin-like substance into renal venous blood in response to angiotensin II. Circulation Res. *27:* suppl. 1, pp. 1–121 (1970).

20 *Muirhead, E.E.; Jones, F., and Stirman, J.A.:* Antihypertensive property in renoprival hypertension of extract from renal medulla. J. Lab. clin. Med. *56:* 167–180 (1960).

21 *Patak, R.V.; Mookerjee, B.K.; Bentzel, C.J.; Hysert, P.E.; Babej, M., and Lee, J.B.:* Antagonism of the effects of furosemide by indomethacin in normal and hypertensive man. Prostaglandins *10:* 649–659 (1975).

22 *Paulsrud, J.R. and Miller, O.N.:* Inhibition of 15-OH prostaglandin dehydrogenase by several diuretic drugs. Fed. Proc. Fed. Am. Socs exp. Biol. *33:* 590 (1974).

23 *Payakkapan, W.; Attallah, A.A.; Lee, J.B.; and Carr, A.A.:* Effect of sodium intake on prostaglandin A, renin and aldosterone in normotensive humans. Kidney int. *8:* S-283–S-290 (1975).

24 *Vander, A.J.:* Direct effects of prostaglandin on renal function and renin release in anesthetized dog. Am. J. Physiol. *214:* 218–221 (1968).

25 *Westura, E.E.; Kannegiesser, H.; O'Toole, J.D., and Lee, J.B.:* Antihypertensive effects of prostaglandin A_1 in essential hypertension. Circulation Res. *27:* suppl. 1, pp. 131–137 (1970).

26 *Zusman, R.M.; Caldwell, B.V.; Mulrow, P.J., and Speroff, L.:* The role of prostaglandin A in the control of sodium homeostasis and blood pressure. Prostaglandins *3:* 679–690 (1973).

27 *Zusman, R.M.; Spector, D.; Caldwell, B.V.; Speroff, L.; Schneider, G., and Mulrow, P.J.:* The effect of chronic sodium loading and sodium restriction on plasma prostaglandin A, E and F concentrations in normal humans. J. clin. Invest. *52:* 1093–1098 (1973).

Dr. *R. Düsing,* Department of Medicine, E.J. Meyer Memorial Hospital,
462 Grider Street, *Buffalo, NY 14215* (USA)

Contr. Nephrol., vol. 12, pp. 54–68 (Karger, Basel 1978)

Prostaglandins and High Blood Pressure

A. Hornych

Centre de Recherches sur l'Hypertension artérielle (Dir. Pr. *P. Milliez*), U-28 Inserm (Dir. Pr. *J. Bariety*), Hôpital Broussais, Paris

Introduction

The kidney plays a double role in blood pressure regulation by secreting (1) vasopressor substances as renin-angiotensin system (40), nephrotensin (13), renopressin (43) and others (5), and (2) vasodepressor substances as lipids: (a) neutral lipids – renomedullary antihypertensive lipid (37), (b) acid lipids – prostaglandins (28), (c) basic lipids – phospholipids (44), and kinins (8).

Lee et al. (29) isolated from renal medullary hypotensive extracts three acid lipids which were identified as prostaglandins A_2, E_2, and $F_{2\alpha}$. PGA_2 and PGE_2 are hypotensive (3, 15, 27, 29, 30), $PGF_{2\alpha}$ is hypertensive in most species including man (38). Since *Lee*'s discovery an important effort has been made to clarify the role of prostaglandins in the pathogenesis of high blood pressure. The present work contributes to this question resuming the results of a systemic investigative study in clinical hypertension.

The progress in this field was limited by the development of accurate methods for the measurement of plasma prostaglandins. For this reason a very sensitive radioimmunoassy method was developed for the measurement of plasma prostaglandins of the A_1, B_1, E_2 and F_α class, able to detect a concentration as low as 2–3 pg/ml and enough specific (7, 20). In spite of several objections raised against the natural existence of PGA_2 (4, 6, 11, 26) we started with the PGA class for three reasons: (1) PGA is not inactivated in pulmonary circulation (9, 33), therefore represents a circulating prostaglandin, (2) PGA varies with low and high Na intake (11, 31, 55), (3) PGA inhibits the renin-angiotensin system (17, 25, 32). Our radioimmunoassay method detects a 'PGA immunoreactive material' which includes PGA_1, to a lesser extent PGA_2 and possibly some first metabolites of PGA_1 and PGA_2, and cross-reacts with PGB_1 (20). At a second time we introduced the radioimmunoassay measurement of PGB_1 (20) and PGE_2 and $PGF_{1,2\alpha}$ with slight modifications of the described methodology (7).

The developed radioimmunoassay methods have been used for the measurement of plasma prostaglandins in different groups of hypertensive patients and control subjects. The present work resumes our experience and demonstrates the significant involvement of prostaglandins in blood pressure regulation.

Subjects

The following groups of subjects have been investigated:

(1) 23 control subjects (normotensive healthy volunteers), mean age of 24 years (21–40 years) on *ad libitum* Na intake; (2) 6 anephric patients maintained by chronic hemodialysis, mean age of 37 years (28–57 years); (3) 9 renal hypertensive patients with bilateral kidney disease in terminal renal failure maintained by chronic hemodialysis, mean age of 39 years (27–64 years); (4) 7 hypertensive patients with chronic bilateral kidney disease without renal failure, mean age of 49 years (24–70 years); (5) 11 patients with chronic bilateral kidney disease with normal arterial blood pressure, mean age of 38 years (14–74 years); (6) 8 patients with renovascular hypertension due to unilateral renal arterial stenosis, mean age of 44 years (18–68 years); (7) 13 labile hypertensive patients, mean age of 23 years (18–28 years); (8) 26 essential hypertensive patients, mean age of 45 years (25–69 years). Prostaglandins have been measured in peripheral blood of fasting subjects sampled in a sitting position in the morning. All patients had the same sodium (90 mEq) and potassium (40–50 mEq) intake. Prostaglandins and plasma renin activity have been measured in vena caval blood beneath the renal veins and in both renal veins of the following groups of patients: (9) 8 hypertensive patients with unilateral renal arterial stenosis, mean age of 42 years (18–55 years); (10) 7 hypertensive patients with unilateral renal atrophy with intact main renal arteries, mean age of 42 years (19–53 years), and (11) 1 patient with normal kidneys.

Methods

Prostaglandin Measurement

20 ml of blood was sampled in chilled plastic tubes containing EDTA 2 mg/ml (Hopkin & Williams, Ltd.) and indomethacin powder 0.1 mg/ml (Merck, Sharp & Dohme). The blood was kept on ice and the plasma was separated by 20 min centrifugation at 4,500 rpm (4 °C) as soon as possible but no later than 60 min after sampling. 3–5 ml of fresh plasma was extracted the same day twice by redistilled ethyl acetate (Merck) separately for the PGA and B class, and another 3–5 ml for PGE_2 and PGF_α according to the described methodology (7, 20). Prostaglandins were separated by silic acid column chromatography and measured by radioimmunoassay with specific antisera raised against PGA_1, B_1, E_2, and $F_{2\alpha}$ (Institut Pasteur). The results are expressed in pg/ml of plasma.

Table I. Plasma prostaglandins A_1, B_1, E_2, and $F_{1,2\alpha}$ in peripheral blood of 20 control subjects (healthy normotensive volunteers)

	Age years	Arterial blood pressure, mm Hg		Prostaglandins pg/ml			
		systolic	diastolic	A_1	B_1	E_2	$F_{1,2\alpha}$
Min.	21	90	60	12	39	0	1.8
Max.	48	140	90	276	308	16.3	41.8
Mean	24	121	76	94	189	5.5	17.2
± SE	2.0	3.2	1.8	15	14	0.8	3.3

Plasma renin activity (PRA) was measured by radioimmunoassay of angiotensin I liberated during 3 h of incubation (36). The results are expressed in ng/ml/h. Statistical analysis was performed by t test and linear regression analysis. All results are expressed by mean value ± SE if not otherwise stated.

Results

The results of measurement of four prostaglandins in peripheral plasma of 20 control normotensive volunteers are summarized in table I.

As concerns prostaglandin A, we verified at first, whether this prostaglandin is not inactivated in pulmonary circulation of man. Indeed, the measurement of this prostaglandin in simultaneously sampled arterial and venous blood in 9 essential hypertensive patients demonstrated that the mean concentration in arterial blood was only 3% lower than in venous blood (fig. 1). Therefore our data are in accordance with published experimental data (9, 33). However, in 2 patients the arterial concentration was higher than the venous one, suggesting the possible pulmonary generation of PGA. This supposition was confirmed by the measurement of PGA in peripheral blood of 6 anephric patients (18). Only in 2 of them PGA was undetectable in peripheral blood. All others had definite levels of PGA in the normal range. Therefore, it is necessary to admit the extrarenal generation of PGA.

On the other hand, this prostaglandin is inactivated in hepatic circulation because the concentration in hepatic venous blood in 4 patients with renovascular hypertension was 30% lower than in vena caval blood (20). These results are concordant with experimental data of Horton and Jones (22).

The PGA immunoreactive material is distinct from PGB because the measurement of the same plasma extract with two different antisera, one more specific for PGA and the other specific for PGB, revealed a much higher proportion of PGA than PGB in pooled plasma samples in both renal veins of a

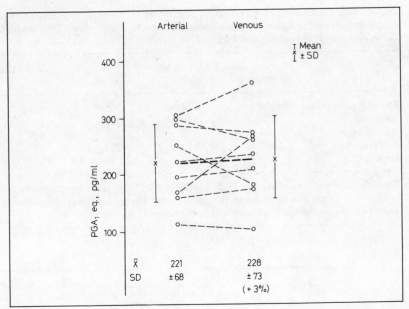

Fig. 1. PGA$_1$ concentration in arterial and venous blood of 9 essential hypertensive patients.

patient with renovascular hypertension (table II). The proportion of PGA was much lower in vena caval blood and especially in hepatic venous blood.

PGA in Peripheral Blood of Different Groups of Hypertensive Patients

The results of measurement of PGA$_1$ in peripheral blood of different groups of hypertensive patients and control subjects are summarized in figure 2.

The mean concentration of PGA$_1$ in peripheral blood of 23 control subjects was 115 ± 15 pg/ml; in 6 anephric patients it was 51 ± 21 pg/ml (n.s.), in 9 hypertensive patients with end-stage kidney disease it was 231 ± 51 pg/ml (p <0.01 in comparison with control subjects), in 7 hypertensive patients with chronic nephropathies without renal failure it was 204 ± 60 mg/ml (p <0.001), in 11 normotensive patients with chronic nephropathies it was 136 ± 30 pg/ml (n.s.), in 8 patients with renovascular hypertension it was 253 ± 20 pg/ml (p <0.001) and in 26 essential hypertensive patients the concentration was 224 ± 16 pg/ml (p <0.001) (19).

Therefore PGA was significantly increased in all hypertensive patients, it was normal only in normotensive nephropathies, decreased in anephric patients, absent in 2 anephric patients and in 1 patient with end-stage kidney disease.

Table II. Concentration of PGA$_1$ and PGB$_1$ immunoreactive material measured in the same plasma extract by the antiserum raised against PGA$_1$ (which cross-reacts 100% with PGB$_1$) and PGB$_1$ (which cross-reacts only 9% with PGA$_1$)

	PGA$_1$ equiv. pg/ml	PGB$_1$ equiv. pg/ml	A–B pg/ml	% of PGA$_1$ equiv.
Pooled plasma sample	580	214	366	63
Vena cava	171	140	31	18
Right renal vein	547	179	368	67
Left renal vein	382	102	280	73
Hepatic vein	144	130	14	9

The difference between PGA$_1$ and PGB$_1$ (probable PGA) is expressed in pg/ml in the third column and in percent of PGA$_1$ in the last column. The measurement was performed in pooled plasma sample and in patients with renovascular hypertension.

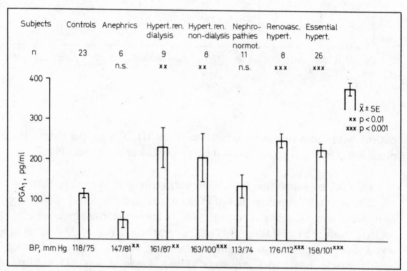

Fig. 2. PGA$_1$ concentration in peripheral blood of control subjects, anephric patients and different groups of hypertensive and normotensive patients.

When we analyzed the relationship between arterial blood pressure, systolic or diastolic, and peripheral PGA$_1$, we found a positive significant correlation (n = 83, r = 44, p <0.001) between these two parameters (fig. 3). This means that higher blood pressure is associated, in general, with increased PGA$_1$ concentration in peripheral blood.

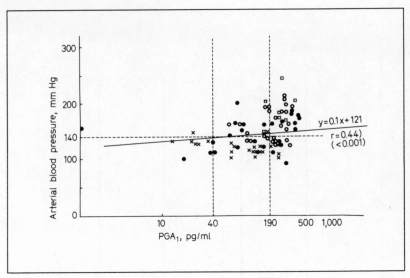

Fig. 3. The correlation between systolic arterial blood pressure and PGA₁ concentration in peripheral blood. X = Controls (23), ○ = essential hypertension (26), □ = renovascular hypertension (7), ● = nephropathies (27).

Table III. Peripheral plasma concentration of PGE_2 and $PGF_{1,2\alpha}$ in 10 essential hypertensive patients and 20 control normotensive subjects

Subjects	n	Mean BP mm Hg	Prostaglandins, pg/ml	
			E_2	$F_{1,2\alpha}$
Controls	20	121/76	5.5 ± 0.8	17.2 ± 3.3
Essential hypertension	10	183/107	59.9 ± 21.0	19.8 ± 5.4
t test. p <		0.001	0.001	n.s.

Essential Hypertension

Figure 2 demonstrates increased levels of PGA₁ in this clinical hypertension. We measured also PGE_2 and PGF_α in a preliminary group of 10 essential hypertensive patients, in peripheral blood only. The results are summarized in table III.

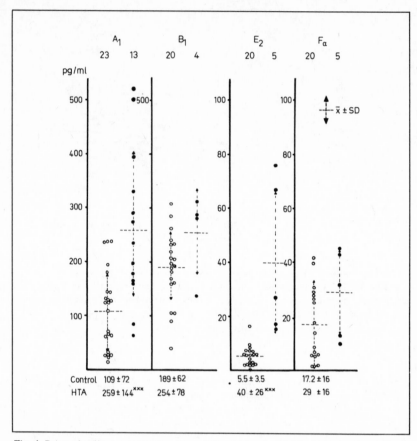

Fig. 4. Prostaglandin concentration in peripheral blood of labile hypertensive patients (•) and control normotensive subjects (○).

PGE_2 was significantly increased in the whole group of patients in comparison with control subjects (p <0.001). This prostaglandin was in a normal range in only 3 out of 10 patients and was present in peripheral plasma of all subjects. PGF_α was not significantly different from control subjects and only in 3 of them was it increased. It was detectable in all subjects.

Labile Hypertension

PGA was measured in peripheral blood of 13 labile hypertensive subjects. Other prostaglandins were measured only in 5 of them. The results are summarized in figure 4. PGA_1 an PGE_2 were significantly increased in labile hypertension compared with control subjects (p <0.001); PGB_1 and PGF_α were not significantly modified (21). Therefore it seems that prostaglandin levels are similarly changed in labile as in essential hypertension patients.

Renovascular and Renal Hypertension

Plasma prostaglandins and PRA have been measured in vena caval blood beneath renal veins (VC), in renal venous blood from 'normal kidney' (VRN) and in the blood from pathological kidneys (VRP) either with renal arterial stenosis or atrophy in 8 patients with unilateral renal arterial stenosis and for comparison in 7 patients with unilateral renal atrophy (with intact main renal arteries). The results have been compared with normal peripheral concentration of different prostaglandins (table I) and peripheral PRA measured in 20 control subjects in supine position, which was 0.89 ± 0.05 ng/ml/h. The ratio of renal venous prostaglandin concentration and peripheral venous concentration has been measured in a patient with normal kidneys so as to know if it is possible to compare the renal venous concentration of prostaglandins of hypertensive patients with peripheral venous concentration of prostaglandins of control subjects. This ratio, expressed in percent, was 132% for PGA, 135% for PGB, 110–136% for PGE_2 and 95–116% for PGF_α. This means that the renal venous concentration of prostaglandins was higher than in peripheral venous blood in a similar proportion to that described for PRA by *Sealey et al.* (42), except PGF, which seems to be lower. The results of the measurements are summarized in figure 5.

Renovascular Hypertension

The mean arterial blood pressure 193/122 mm Hg was significantly higher than in control subjects (p $<$0.001). Mean PRA in vena caval (VC), normal renal vein (VRN) and in the renal venous blood from the kidney with stenosis (VRP) were, respectively, 3.94 ± 1.27, 4.89 ± 1.66 and 9.07 ± 4.28 ng/ml/h. All these values are significantly higher than in control subjects (p $<$0.001). Prostaglandins in VC, VRN, and VRP PGA_1 were 305 ± 90, 341 ± 50 and 319 ± 76 pg/ml. All these values are significantly higher than in control subjects (p $<$0.001). The concentration in the blood from the normal kidney (with higher perfusion pressure, not protected by stenosis) was a little bit higher than from the kidney with stenosis. PGB_1: 209 ± 45, 175 ± 51, 185 ± 85 pg/ml. These values are not different from control subjects. PGE_2: 15.0 ± 5.3, 18.5 ± 7.4, 37.8 ± 18.0 pg/ml. All these values are significantly higher than in control subjects (p $<$0.001). The highest concentration was found on the stenotic side but in 2 out of 8 patients, PGE_2 was undetectable in renal venous blood of the stenotic side. PGF_α: 16.9 ± 4.8, 26.5 ± 9.5, 33.3 ± 10.9 pg/ml. The highest concentration was found on the stenotic side where the increase approaches the statistical significance (p $<$0.1 $>$0.05). Other values are not significantly different from control subjects.

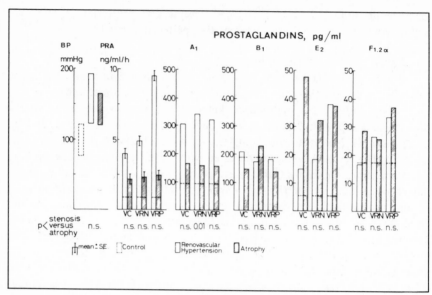

Fig. 5. The results of measurement of prostaglandin concentration and plasma renin activity (PRA) in vena caval blood beneath renal veins (VC), in renal venous blood of 'normal' kidney (VRN) and in renal venous blood of pathologic kidney (VRP) either with stenosis or atrophy in 8 patients with unilateral renal arterial stenosis – renovascular hypertension, and in 7 patients with unilateral renal atrophy – renal hypertension. The interrupted line represents the mean concentration of PRA and prostaglandins in peripheral blood of 20 control subjects. For the results of statistical comparison between hypertensive patients and control subjects see text.

Renal Hypertension

The mean arterial blood pressure of these patients 167/112 mm Hg was also significantly higher than in control subjects (p <0.001). In contrast to the former group, PRA and PGA were only moderately increased without gradient between kidneys and were not significantly different from control subjects. The respective concentrations in VC, VRN and VRP blood were: PRA: 1.63 ± 0.91, 1.7 ± 1.04, 1.92 ± 0.92 ng/ml/h; PGA_1: 154 ± 55, 144 ± 32, 151 ± 35 pg/ml; PGB_1: 140 ± 41, 222 ± 64, 139 ± 38 pg/ml. Conversely, the increase of PGE_2 and PGF_α was greater than in patients with renal arterial stenosis. The respective values in VC, VRN and VRP blood were: PGE_2: 46.5 ± 20.5, 27.2 ± 13.1, 35.8 ± 17.4 pg/ml. All these values are significantly higher than in control subjects (p <0.001). The highest concentration of PGE_2 was found in vena caval blood. This result suggests the possible extrarenal synthesis or decreased metabolism of this prostaglandin, or the participation of both mechanisms. PGE_2 was not detectable in renal venous blood of atrophic kidney in 1 out of 7 patients.

PGF_α: 28.8 ± 9.4, 24.3 ± 9.3, 37.4 ± 11.9 pg/ml. The highest concentration of PGF_α was found in the blood·from atrophic kidney and this increase was statistically significant in comparison with control subjects ($p < 0.05$).

The analysis of the relationship between PRA and different prostaglandins in all simultaneously sampled blood, i.e. from vena cava or both renal veins, in both groups of patients (stenosis, atrophy) revealed a highly significant positive correlation between PRA and PGA_1 ($n = 44$, $r = 0.51$, $p < 0.001$) and PGB_1 ($n = 38$, $r = 0.59$, $p < 0.001$) but not with PGE_2 or PGF_α.

Concerning the renovascular and renal hypertension it seems that the renovascular component interacts more with the renin-angiotensin-PGA system and that the renal parenchymatous component interacts more with PGE_2 and PGF_α.

Discussion

The presented results demonstrate the clinical importance of PGA immunoreactive material in plasma of man in spite of several objections raised against the natural existence of PGA_2 (4, 6, 11, 26). Different groups of workers described the presence of PGA in human plasma (1, 23, 24, 53, 54) and also the pathophysiological implications of this prostaglandin in (a) low and high sodium intake (11, 31, 55), (b) blood pressure lowering renal tumor (56), and (c) Bartter's syndrome (10, 45, 46). Our data in hypertensive subjects corroborate the former and the latter observations. We have found: (1) increased levels of PGA in several groups of hypertensive patients and not in normotensive nephropathies, (2) a positive significant correlation between arterial blood pressure and PGA, (3) a positive significant correlation between PRA and PGA, (4) increased levels of PGA not only on the stenotic side of patients with renal arterial stenosis but also in renal venous blood of the 'normal kidney' with higher perfusion pressure. Therefore we conlude:

I. The increase of PGA may represent: (1) a secondary antihypertensive mechanism — in accordance with the hypothesis of *Tobian et al.* (50) who demonstrated that increased renal perfusion pressure releases vasodepressor lipids from the kidney; (2) a secondary diuretic and natriuretic mechanism — in accordance with the hypothesis of *Guyton et al.* (14) who demonstrated that increased renal perfusion pressure increases the natriuresis.

Therefore PGA may represent a mediator of 'pressure natriuresis', especially in the contralateral kidney of patients with unilateral renal arterial stenosis. As PGA is not inactivated in pulmonary circulation, it may represent a circulating 'antihypertensive hormone'.

These conclusions could be extended to PGE_2, which is also hypotensive, diuretic and natriuretic in man (23, 38) but with some limitations: PGE_2, in contrast to PGA, does not inhibit the generation of angiotensin I (25); there is no correlation between PRA and PGE_2, therefore this prostaglandin is less effective in the inhibition of the renin-angiotensin system. Finally, it is inactivated in pulmonary circulation (9, 33), unless the renal or pulmonary degradation is depressed, as suggested by our results in renal hypertension. Then this prostaglandin may function also as circulating prostaglandin.

The release of PGA_1 and PGE_2 on the stenotic side with decreased perfusion pressure (and to some extent also in the contralateral kidney) is mediated by other mechanisms which are ischemia (35) and/or angiotensin (16, 34, 48, 49).

II. The decrease or absence of PGE_2 in renal venous blood of 3 hypertensive patients out of 15 could be of clinical significance. The inadequate synthesis of hypotensive prostaglandins may contribute to the prevalence of vasoconstrictor-hypertensive systems and by this way contribute to the increase of blood pressure.

The validity of these hypotheses about the hypotensive, diuretic and natriuretic role of prostaglandin A and E_2 is supported by two clinical evidences: (1) Bartter's syndrome is a normotensive syndrome sometimes accompanied by orthostatic hypotension in spite of considerable stimulation of the renin-angiotensin-aldosterone system (2, 47, 52). This syndrome is associated with an excess of PGA (10, 45, 46) and PGE_2 (12, 46). (2) The inhibition of prostaglandin synthesis by indomethacin increases the arterial blood pressure in man (39).

III. The increase of PGF_α in renal venous blood from the kidney with renal arterial stenosis or atrophy may represent a direct hypertensive mechanism. Therefore, PGF could be one of the renal factors responsible for the increase of blood pressure, especially when PRA is normal, as we observed in patients with renal atrophy. The importance of this prostaglandin may be enhanced by decreased pulmonary degradation allowing the systemic effect. Increased concentration of PGF in vena caval blood of patients with renal atrophy seems to corroborate this possibility. The possible role of PGF in the pathogenesis of high blood pressure is suggested also by some experimental data (41, 51).

In conclusion, the presented work demonstrates the significant involvement of prostaglandins in the pathogenesis of clinical hypertension because they interfere with (a) the renin-angiotensin system, (b) volume-blood pressure regulation, and (c) general and regional hemodynamics. The inadequate synthesis of hypotensive prostaglandis or an excess of hypertensive PGF may contribute to the increase of blood pressure.

Summary

Prostaglandins A_1, B_1, E_2, F_α and PRA have been measured by radioimmunoassay in peripheral or renal venous blood of different groups of hypertensive and control subjects. PGA_1 and PGE_2 were significantly increased in renal, renovascular, labile and essential hypertension. PGF_α was significantly increased only in patients with unilateral renal atrophy and in some patients with renovascular and essential hypertension. There was a significant positive correlation between PRA and PGA_1 or B_1, but not with PGE_2 or F_α. The increase of PGA and PGE represents a secondary antihypertensive, diuretic and natriuretic mechanism, the increase of PGF a direct hypertensive mechanism.

Acknowledgements

The author wishes to express his thanks to all co-authors of separate works who enabled this revue, mainly Prof. *P. Milliez*, Prof. *J. Bariety*, Prof. *M. Safar*, Dr. *F. Fontaliran*, Dr. *Y. Weiss*, Prof. *P. Corvol*, Prof. *J. Menard*, Dr. *T.T. Guyene;* to Dr. *J.E. Pike* of the Upjohn Company, Kalamazoo, Mich., for the generous gift of prostaglandin standards; to Mrs. *D. Terana* for technical assistance, and to Inserm for the grants No. 73.5.475.17, 74.5.221.5 and 32.76.64.

References

1   *Attalah, A.A. and Lee, J.B.:* Radioimmunoassay of prostaglandin A. Intrarenal PGA_2 as a factor mediating saline induced natriuresis. Circulation Res. *33:* 696–703 (1973).

2   *Bartter, F.C.; Pronove, P.; Gill, J.R.; MacCardle, C.R., and Diller, E.:* Hyperplasia of the juxtaglomerular complex in the hyperaldosteronism and hypokalemic alkalosis. Am. J. Med. *33:* 811–828 (1962).

3   *Bergström, S.; Dunner, H.; Euler, U.S. von; Pernow, B., and Sjovall, J.:* Observations on the effects of infusion of prostaglandin-E in man. Acta physiol. scand. *45:* 145–151 (1959).

4   *Crowshaw, K.:* The incorporation of $(1-^{14}C)$ arachidonic acid into the lipids of rabbit renal slices and conversion to prostaglandin E2 and F2. Prostaglandins *3:* 607–620 (1973).

5   *Croxatto, H.R.:* Plasma of serum vasopressor peptides other than angiotensins; in *Page and Bumpus* Angiotensin, pp. 240–259 (Springer, Berlin 1974).

6   *Daniels, E.G.; Hinman, J.W.; Leach, B.E., and Muirhead, E.E.:* Identification of prostaglandin E_2 as the principal lipid of rabbit renal medulla. Nature, Lond. *215:* 1298–1299 (1967).

7   *Dray, F.; Charbonel, B., and Maclouf, J.:* Radioimmunoassay of prostaglandins F alpha, E1 and E2 in human plasma. Eur. J. clin. Invest. *5:* 311–318 (1975).

8   *Erdös, E.G.:* Handbook of experimental pharmacology, vol. 25: Bradykinin, kallidin and kallikreins (Springer, Berlin 1970).

9   *Ferreira, J.H. and Vane, J.R.:* Prostaglandins: Their disappearance from and release into the circulation. Nature, Lond. *216:* 868–873 (1967).

10   *Fichman, M.D.; Telfer, N.; Zia, P.; Speckart, D.; Golub, M., and Rude, R.:* Role of prostaglandins in the pathogenesis of Bartter's syndrome. Am. J. Med. *60:* 785–797 (1976).

11   *Frolich, J.C.; Sweetman, B.J.; Carr, A.A.; Hollifield, V.W., and Oates, J.A.:* Assessment of the bevels of PGA2 in human plasma by gas-chromatography – mass spectrometry. Prostaglandins *10:* 185–195 (1975).

12   *Gill, S.R.; Frölich, J.C.; Bowden, R.E.; Taylor, A.A.; Keiser, H.R.; Seyberth, H.W.; Oates, J.A., and Bartter, F.C.:* Bartter's syndrome: A disorder characterized by high

urinary prostaglandins and a dependence of hyperreninemia on prostaglandin synthesis. Am. J. Med. *61:* 43–51 (1976).

13 *Grollman, A. and Krishnamurty, V.S.R.:* A new pressor agent of renal origin: its differentiation from renin and angiotensin. Am. J. Physiol. *221:* 1499–1506 (1971).

14 *Guyton, A.C.; Coleman, T.G.; Cowley, A.W.; Manning, R.D.; Norman, R.A., and Ferguson, J.D.:* A system analysis approach for understanding long-range arterial blood pressure control and hypertension. Circulation Res. *35:* 159–176 (1974).

15 *Hornych, A.; Safar, M.; Papanicolaou, N.; Meyer, P., and Milliez, P.:* Renal and cardiovascular effect of prostaglandin A_2 in hypertensive patients. Eur. J. clin. Invest. *3:* 391–398 (1973).

16 *Hornych, A. and Papanicolaou, N.:* Prostaglandins in renal venous blood of essential hypertensive patients. Prostaglandins *7:* 383–386 (1974).

17 *Hornych, A. et Papanicolaou, N.:* Effet de la prostaglandine A_2 chez les malades insuffisants rénaux. Nouv. Presse méd. *3:* 2628–2632 (1974).

18 *Hornych, A.; Bedrossian, J.; Bariety, J.; Menard, J.; Corvol, P.; Safar, M.; Fontaliran, F., and Milliez, P.:* Prostaglandins and hypertension in chronic renal disease. Clin. Nephrol. *4:* 144–151 (1975).

19 *Hornych, A.; Safar, M.; Bariety, J. et Milliez, P.:* Prostaglandines et circulation rénale chez les sujets hypertendus. J. Urol. Néphrol. *82:* 804–811 (1976).

20 *Hornych, A.; Weiss, Y.; Menard, J.; Corvol, P.; Fontaliran, F.; Bariety, J., and Milliez, P.:* Radioimmunoassay of prostaglandins A and B in human blood. Prostaglandins *12:* 383–397 (1976).

21 *Hornych, A.; Safar, M.; London, G.; Weiss, Y.; Bariety, J. et Milliez, P.:* Hypertension artérielle labile et les prostaglandines. J. Urol. Néphrol. *83:* 289–293 (1977).

22 *Horton, E.W. and Jones, L.:* Prostaglandin A_1, A_2 and 19-hydroxy A_1, their action on smooth muscle and their inactivation on passage through the pulmonary and hepatic portal vascular blood. Br. J. Pharmacol. *37:* 705–722 (1969).

23 *Jaffe, B.M.; Behrman, H.R., and Parker, C.W.:* Radioimmunoassay measurement of prostaglandins E, A and F in human plasma. J. clin. Invest. *52:* 398–405 (1973).

24 *Jubiz, W.; Frailey, J.; Child, C., and Bartholomew, K.:* Physiologic role of prosta-glandins of the E (PGE), F (PGF) and AB (PGAB) group. Estimation by radioim-munoassay in unextracted human plasma. Prostaglandins *2:* 471–489 (1972).

25 *Kotchen, J.A. and Miller, M.C.:* Effect of prostaglandins on renin activity. Am. J. Physiol. *226:* 314–318 (1974).

26 *Larsson, C. and Anggård, E.:* Mass spectrometric determination of prostaglandin E_2, $F_{2\alpha}$ and A_2 in the cortex and medulla of the rabbit kidney. J. Pharm. Pharmac. *28:* 326–328 (1976).

27 *Lee, J.B.:* Chemical and physiological properties of renal prostaglandins: the antihyper-tensive effects of medullin in essential human hypertension; in *Bergström and Sammelsson* Prostaglandins. Proc. 2nd Nobel Symp., Stockholm 1966, pp. 197–210 (Interscience, London 1967).

28 *Lee, J.B.; Covino, B.G.; Takman, B.H., and Smith, E.R.:* Renomedullary vasodepressor substance, medullin: isolation, chemical characterisation and physiological properties. Circulation Res. *17:* 57–77 (1965).

29 *Lee, J.B.; Crowshaw, K.; Takman, B.H.; Attrep, K.A., and Gougoutas, J.Z.:* The identification of prostaglandins E_2, F_2 and A_2 from rabbit kidney medulla. Biochem. J. *105:* 1251–1260 (1967).

30 *Lee, J.B.; McGiff, J.C.; Kannegiesser, H.; Aykent, J.Y.; Mudd, J., and Frawley, F.F.:* Prostaglandin A_1: antihypertensive and renal effects. Studies in patients with essential hypertension. Ann. intern. Med. *74:* 703–710 (1971).

31 *Lee, J.B.; Patak, P.V., and Mookerjee, B.K.:* Renal prostaglandins and the regulation of blood pressure and water homeostasis. Am. J. Med. *60:* 798–816 (1976).

32 *McClatchey, W.M. and Carr, A.A.:* Prostaglandin (PGA$_1$) angiotensin and renal function. Prostaglandins *2:* 213–217 (1972).

33 *McGiff, J.G.; Terragno, N.A.; Strand, J.C.; Lee, J.B.; Lonigro, A.J., and Ng, K.K.F.:* Selective passage of prostaglandins across the lung. Nature, Lond. *223:* 742–745 (1969).

34 *McGiff, J.C.; Crowshaw, K.; Terragno, N.A., and Lonigro, A.J.:* Release of prostaglandin-like substance into renal venous blood in response to angiotensin II. Circulation Res. *27:* suppl. I, pp. 121–130 (1970).

35 *McGiff, J.C.; Crowshaw, K.; Terragno, N.A.; Lonigro, A.S.; Strand, J.C.; Williamson, M.A.; Lee, J.B., and Ng, K.K.F.:* Prostaglandin-like substances appearing in canine renal venous blood during renal ischemia. Circulation Res. *27:* 765–782 (1970).

36 *Menard, J. and Catt, K.J.:* Measurement of renin activity, concentration and substrate in rat plasma by radioimmunoassay of angiotensin I. Endocrinology *90:* 422–430 (1972).

37 *Muirhead, E.E.; Leach, B.E.; Byers, L.W.; Brooks, B.; Daniels, E.G., and Hinman, J.W.:* Antihypertensive neutral renomedullary lipids (ANRL); in *Fischer* Kidney hormones, pp. 486–503 (Academic Press, London 1971).

38 *Nakano, J.:* General pharmacology of prostaglandins; in *Cuthbert* The prostaglandins: pharmacological and therapeutic advances, pp. 23–124 (Heineman, London 1973).

39 *Patak, R.U.; Mookerjee, B.K.; Bentzel, C.J.; Hysert, P.E.; Babej, M., and Lee, J.B.:* Antagonism of the effects of furosemide by indomethacin in normal and hypertensive man. Prostaglandins *10:* 649–659 (1975).

40 *Peart, W.S.:* Renin angiotensin system in hypertensive disease; in *Fischer* Kidney hormones, pp. 217–238 (Academic Press, London 1971).

41 *Rønne, J.A. and Arrigoni-Martelli, E.:* Renal prostaglandin metabolism in spontaneously hypertensive rats. Biochem. Pharmac. *26:* 485–488 (1977).

42 *Sealey, J.E.; Bühler, F.R.; Laragh, J.H., and Waughan, E.D.:* The physiology of renin secretion in essential hypertension. Estimation of renin secretion rate and renal plasma flow from peripheral and renal vein renin levels. Am. J. Med. *55:* 391–400 (1973).

43 *Skeggs, L.T.; Kahn, J.R.; Levine, M.; Dorer, F.E., and Lentz, K.E.:* Chronic one-kidney hypertension in rabbits. III. Renopressin, a new hypertensive substance. Circulation Res. *40:* 143–149 (1977).

44 *Smeby, R.R. and Bumpus, F.M.:* Renin inhibitors; in *Fisher* Kidney hormones, pp. 207–216 (Academic Press, London 1971).

45 *Smiley, J.W.; Del Guercio, E.T.; March, N.M., and Hornych, A.:* Acute effect of oral indomethacin on renal vein prostaglandin and renin levels in a patient with Bartter's syndrome. Abstract. Kidney int. *10:* 511 (1976).

46 *Smiley, J.W.; Del Guercio, E.J.; March, N.M. et Hornych, A.:* Syndrome de Bartter. Syndrome d'hyperprostaglandinisme primaire rénal. J. Urol. *83:* 693–699 (1977).

47 *Solomon, R.J. and Brown, R.S.:* Bartter's syndrome. New insights into pathogenesis and treatment. Am. J. Med. *59:* 575–583 (1975).

48 *Speckart, P.; Golub, M.; Zia, P.; Zipser, R., and Horton, R.:* The effect of angiotensin II and indomethacin on immunoreactive prostaglandin 'A' levels in man. Prostaglandins *11:* 481–488 (1976).

49 *Terragno, D.A.; Strand, J.C.; Pacholczyk, D.A., and McGiff, J.C.:* Prostaglandin E$_2$, an intrarenal hormone; in Prostaglandins, pp. 207–224 (Inserm, Paris 1973).

50 *Tobian, L.; Schonnings, J., and Seefeldt, C.:* The influence of arterial pressure on the antihypertensive action of a normal kidney, a biological servomechanism. Ann. intern. Med. *60:* 377–383 (1964).

51 *Tobian, L. and O'Donnell, M.:* Renal prostaglandins in relation to sodium regulation and hypertension. Fed. Proc. Fed. Am. Socs exp. Biol. *35:* 2388–2392 (1976).

52 *White, M.G.:* Bartter's syndrome. Archs intern. Med. *129:* 41–47 (1972).

53 *Zia, P.; Golub, M., and Horton, R.:* Radioimmunoassay for prostaglandin A_1 in human peripheral blood. J. clin. Endocr. Metab. *41:* 245–252 (1975).

54 *Zusman, P.; Caldwell, B.F., and Speroff, L.:* Radioimmunoassay of the A prostaglandins. Prostaglandins *2:* 41–53 (1972).

55 *Zusman, R.M.; Spector, D.; Caldwell, B.V.; Speroff, L.; Schneider, G., and Mulrow, P.J.:* The effect of chronic sodium leading and restriction on plasma prostaglandin A, E and F concentrations in normal humans. J. clin. Invest. *52:* 1093–1098 (1973).

56 *Zusman, R.M.; Snider, J.S.; Cline, A.; Caldwell, B.V., and Speroff, L.:* Antihypertensive function of a renal cell carcinoma. New Engl. J. Med. *290:* 843–845 (1974).

Dr. *A. Hornych,* Centre de Recherches sur l'Hypertension artérielle, Hôpital Broussais, 96, rue Didot, *F–75014 Paris* (France)

Contr. Nephrol., vol. 12, pp. 69–81 (Karger, Basel 1978)

Case for a Renomedullary Blood Pressure Lowering Hormone

E. Eric Muirhead

University of Tennessee, Center for the Health Sciences and Baptist Memorial Hospital, Memphis, Tenn.

The emergence of endocrinology as a bioscience resulted from key but diverse observations, beginning about 130 years ago (1, 3, 8, 16). These observations, in time, gave rise to a schema considered necessary in order to establish a new endocrine system. In this presentation, certain major features of such schema that support the existence of an antihypertensive endocrine function of the kidney, and especially the renal medulla and its interstitial cells (the renomedullary interstitial cells or RIC), are emphasized. Of necessity, such approach invokes a putative antihypertensive renomedullary hormone (ARH).

It is proposed to identify components of the above-mentioned schema, relate an abridged historical perspective concerning the component and proceed to demonstrate by available experimental data how the kidney and its RIC system appear to be antihypertensive via an endocrine mechanism.

Absence of a Tissue (at the Outset a Ductless Gland, Later Specific Cellular Areas) Is Associated with an Abnormality or Group of Abnormalities

Historical Perspective (Abridged)
Thomas Addison (1855) described a syndrome of man, later bearing his name, and by careful clinicopathologic observations ascribed it to the absence (destruction) of the suprarenal capsules. It is likely that this key observation led to the ablation approach. *Gull* (1873) described the syndrome later termed myxedema, and *Horseley* (1888) reproduced it in monkeys by ablation of the thyroid gland. *Von Mering and Minkowski* (1889) ablated the pancreas of the dog and created experimental diabetes mellitus. *MacCallum and Voegtlin* (1909) induced hypocalcemic tetany of the rabbit by ablation of the parathyroid glands and related it to calcium metabolism.

Case for Kidney and Renal Medulla

The ablation approach seems more convincing for a relationship of the kidney to the hypertensive state if considered in association with other features of the schema, especially transplantation and other renal manipulations. Removal of both kidneys of the rat (4), dog (9) and rabbit (23) makes the animal susceptible to hypertension when an increase in body fluid transpires (renoprival hypertension). The increase in body fluid can be by either an input of Na and water (14) or whole blood (7). Deviation of urine flow into a vein (ureterovenous anastomosis) prevents this hypertensive state (9, 12, 29) under conditions where fluid expansion is equal to that following renal ablation. Ureteral ligation plus a volume load does not prevent the hypertensive state, and under these conditions, the renal medulla (papilla) is destroyed (20, 37). Chemical destruction of the renal medulla enhances the hypertensive state (11). It seems that ablation of the renal medulla allows prohypertensive mechanisms to operate under improper control.

Fibrosis of the renal medulla, as observed in autopsy material, correlates significantly with the existence of hypertension during life and the presence of renal arterial and arteriolar sclerosis (10). In this condition, the environment of the RIC is changed from that of acid mucopolysaccharide (glucoseaminoglycans) to one of collagen. Such a morphologic alteration could cause a dysfunction on the part of these cells.

In malignant hypertension of man the RIC are decreased in number (19). In partial nephrectomy-salt hypertension of the rat the RIC are not only decreased in number but are degenerated in appearance (38).

Transplantation of Suspected Tissue (or Cells) from an Appropriate Source either Prevents or Reverses Putative Endocrine Abnormality

Historical Perspective (Abridged)

A most significant set of observations in this respect was made by *A.A. Berthold* of Göttingen in 1848 (2). *Berthold* asked the question, Why the capon? – and answered it as follows. Two groups of young roosters were studied. One had orchiectomies and became capons (the control group). The other had orchiectomies followed by the introduction of one of the removed testicles into the peritoneal cavity (the test group). The latter group remained muscular, retained their cock's comb, fought other roosters and chased hens. At autopsy, *Berthold* noted a vascularized nodule attached to the peritoneum, seemingly without nervous connections, and concluded 'the testicles act upon the blood and the blood acts correspondingly upon the entire organism'. These experiments were performed 2 years before *Leydig* described the interstitial cells known by his name (8).

The autotransplantation approach was also used to prevent hypothyroidism and Addison's disease (*Schiff*, 1884). The transplantation route established the existence of sex hormones (*E. Knauer*, 1911).

Case for Renal Medulla and Its RIC

It was first demonstrated that either whole kidney transplants (36) or renal perfusion (13) reversed renoprival hypertension of the dog and that this reversal occurred despite the maintenance of a high Na-volume load. Then it was shown that autotransplantation of the renal medulla prevented renoprival hypertension of the dog (35). Although we were not aware of it at the time, in essence, the Berthold maneuver was used in this experiment. Next, it was shown that similar transplants reversed Goldblatt (24) and angiotensin salt hypertension (26) of the rat and prevented malignant (22) and extreme Na-loading (23) hypertension of the rabbit. The antihypertensive action of similar medullary transplants was confirmed in at least five other laboratories (17, 18, 40, 42, 43).

Next, an antihypertensive action of RIC after growth as a monolayer tissue culture (27) was demonstrated toward angiotensin-salt, ony kidney and two kidney Goldblatt (26) and partial nephrectomy-salt hypertensive rats (34). The antihypertensive action of these cells was confirmed in at least two other laboratories (5, 6) (fig. 1).

Evidence that a Substance Formed by Suspected Tissue (or Cells) Is Transported by the Blood Stream in Order to Act on Another Tissue, Often Remote

Historical Perspective (Abridged)

Berthold's suggestion that testicular function was transmitted elsewhere by the blood remained buried in an inaccessible journal until disinterred by *Biedl* in 1910. Credit, therefore, goes to *C. Bernard* (1850s) for presenting the term 'internal secretion' on a sound basis. This he did, however, not for an endocrine substance but for the transfer of glucose from the liver to the blood after its breakdown from glycogen. *Bayliss and Starling* (1905) placed the endocrine concept on sure footing by a set of classic experiments. The pancreas was disconnected from the nervous system, acid chyme introduced into the duodenum caused pancreatic juice to flow. An extract of intestinal mucosa plus acid injected intravenously caused pancreatic juice to flow. Therefore, acid chyme stimulated the mucosa to produce a substance ('secretin') which was secreted into the blood, carried to the pancreas and excited the pancreas to secrete its juice. The substance was termed a hormone, 'to arouse to activity'.

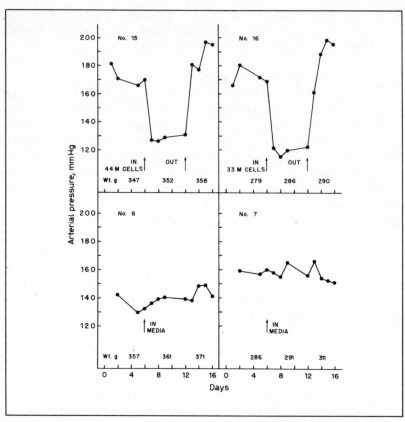

Fig. 1. Transplant of isogeneic renomedullary interstitial cells (Tr TCric) in angiotensin-salt hypertensive (ASH) rats. Following introduction subcutaneously of 44 M and 33 M cells, respectively, there was a sharp lowering of the arterial pressure to normal levels for this colony within 24 h. The pressure remained so depressed until the transplants were removed. Then it returned to prior hypertensive levels. Paired controls receiving tissue culture media had no change in pressure. Most importantly, there was no change in body weight. (Recent, 1977, unpublished data.)

Case for Renal Medulla and Its RIC

The transplants of fragmented renal medulla that exert an antihypertensive action in time consist mostly of RIC (22). Moreover, the areas of RIC are highly vascular and the RIC themselves assume a close relationship to capillaries. The same can be said of transplants of cultured RIC; they also display a prominent positive tropism toward capillaries (26). The arrangement is reminiscent of established endocrine glands. The antihypertensive action of these transplants and their RIC almost certainly result from the secretion of substances into the

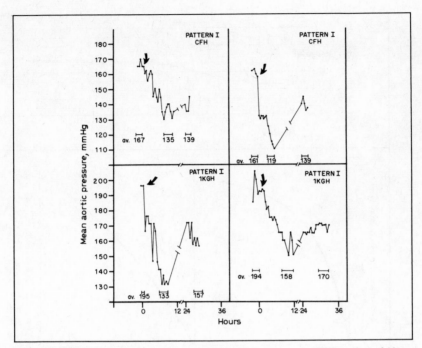

Fig. 2. Transplants of cultured RIC in partial nephrectomy-salt (CFH) and one kidney Goldblatt (1KGH) hypertension. These are individual results. The transplant (∿ 30 M viable cells) was introduced at the arrow. The arterial pressure was lowered to a maximum in a steady manner by 6–12 h. The figures under the bars are averages for the interval of the bar. From Lab. Invest. *35:* 162 (1977).

blood for transport elsewhere. This point was strengthened when it was shown that transplants of cultured RIC exert a significant antihypertensive action within 6–24 h after their introduction; that is, before vascularization has occurred (34). Thus, it appears more plausible to consider that antihypertensive substances seep out of the transplant to enter capillaries and/or lymphatics than to consider that an avascular structure with no intrinsic blood flow neutralizes circulating pressor agents *in situ* (fig. 2).

Extraction of Suspected Tissue (or Cells) Yields Active Principle that Replaces the Function of Tissue (or Cells) when Introduced into the Body

Historical Perspective (Abridged)
Among the first in this area was *G.R. Murray* (1891) who prepared a glycerine extract of the thyroid gland and demonstrated its curative properties

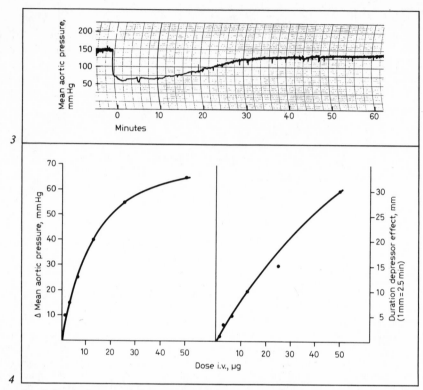

Fig. 3. Acute depressor effect following intravenous bolus dose of renomedullary lipid (RM lipid) in one kidney Goldblatt hypertension, $^1/_{200}$ of fraction. Note that after 60 min the pressure had not yet returned to the original level. (New data.)

Fig. 4. Dose response curve to RM lipid in 1KGH rat. On the left is shown the magnitude of the immediate depressor effect as shown in figure 3. On the right is the duration of the effect. This is a rather crude preparation but the dose effect is demonstrated. At the highest dose the immediate depression amounted to 60 mm Hg and lasted 75 min. (New data.)

toward myxedema. *Schafer and Oliver* (1894) derived an extract of adrenal medulla and showed that it evoked a pressor response. Extracts of the posterior pituitary led to the discovery of oxytocin (*Dale,* 1909) and the antidiuretic hormone (*Farini and von den Velden,* 1913). *Bayliss and Starling* used extracts in support of secretin. *Collip* discovered parathormone (1915–1925) by the extraction route. Then came the superb discovery of insulin by *Banting and Best* (1921) using a modification of the extraction procedure. In the latter two experiments, ablation produced the abnormality, the extract reversed it.

Case for Renal Medulla and Its RIC

A crude extract of fresh renal medulla prevented renoprival hypertension of the dog (30) in a manner identical to that afforded by transplants of renal medulla. This extract also caused an acute as well as a lasting depressor effect when given in larger doses intravenously (30). The active principle was subsequently shown to be a lipid (28). This antihypertensive renomedullary lipid (ARL) (32) reversed renal hypertension of the dog (21) and rat (32) and prevented renoprival hypertension of the dog (25). It was considered to be a neutral moiety and therefore was termed the antihypertensive neutral renomedullary lipid (ANRL) (31) (fig. 3, 4). More recently, a modification of the extraction-purification procedure (34) has yielded from the renal medulla a lipid capable of causing a precipitous acute depressor effect when injected as a bolus dose intravenously as well as a prolonged depressor effect when injected in multiple doses (33, 34). Moreover, a similar lipid extract of cultured RIC also causes the longer lasting depressor effect. The latter extract reproduces the antihypertensive action of transplants of the same cultured RIC (34). Therefore, lipid extracts of renal medulla and its RIC appear to perform the same antihypertensive action as the tissue itself and its isolated cells (RIC) (fig. 5, 6).

Isolation and Identification of Pure Compound from Extracted Active Principle

The ARL, although highly purified, has not been isolated in pure form and its structure remains unidentified.

Measurement of Putative Hormone in Blood and Possibly Other Body Fluids

The putative ARH has not been identified in body fluids. In time, it should be possible to determine whether it exists in body fluids.

Hyperplasia, Adenomatous Change and Neoplasia of Endocrine Tissue with or without Evidence of Hyperfunction

Historical Perspective (Abridged)

An early example of enlargement of the endocrine structure (thyroid gland) and hyperfunction was that of Graves' disease. Outstanding examples became evident: acromegaly (Marie's syndrome) and enlargement of the anterior pituitary, hypercalcemia and hyperparathyroidism and hypoglycemia and insulin-secreting tumors.

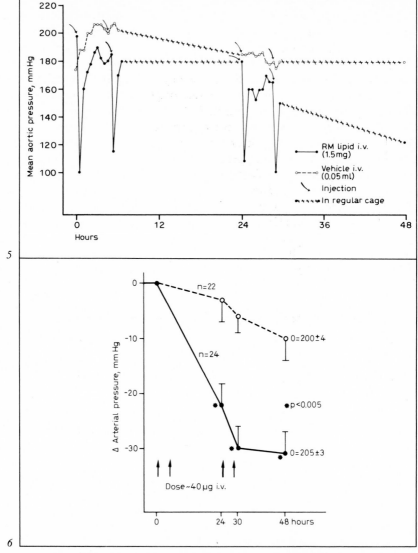

Fig. 5. Depressor effect as well as longer lasting effect of RM lipid injected at arrows (solid line). Recovery from the depressor effect required 1–2 h. By 48 h (∿ 18 h from last dose) the pressure had changed from an original 200 to 120 mm Hg. The paired control receiving vehicle (broken line) had no change in pressure. From *Muirhead et al.* (33).

Fig. 6. Prolonged depressor effect of RM lipid (mean ± SEM). The acute depressor effect that occurred after each dose (upward arrows) was deleted on purpose in order to emphasize the prolonged effect. At 24, 30 and 48 h, the differences between the test and control animals (receiving vehicle) were significant (New data.)

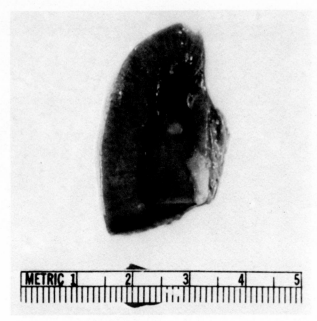

Fig. 7. Renomedullary interstitial cell tumor in its most common location in the human kidney. From *Lerman et al.* (15).

Most endocrine structures are, in some cases, associated with either hyperplasia, adenomatous or carcinomatous change of its functional elements. Such change may or may not be attended by hypersecretion of the corresponding hormone and hyperfunction.

Case for Renal Medulla and Its RIC

Assuming that RIC is an endocrine type of cell, it should undergo hyperplasia or adenomatous change in keeping with the practice of known endocrine organs. With this in mind, *Pitcock and Lerman* suggested that the renomedullary nodule known as the fibroma of the renal medulla or medullary fibrous nodules might represent such a transformation. This turned out to be the case and the nodule was renamed a 'renomedullary interstitial cell tumor' (15). *Prezyna et al.* (39) confirmed this finding. RIC are known to synthesize prostaglandin and the interstitial cell tumor has been shown to be rich in these compounds. However, there is as yet no indication of hyperfunction by the tumor (fig. 7, 8).

In an autopsy service having a high proportion of cardiovascular disease associated with Bright's complex (hypertrophied left ventricle and granular

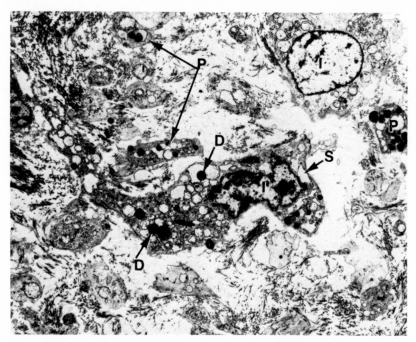

Fig. 8 The main cells of the tumor of figure 7 have the characteristics of the RIC by EM. Note the cytoplasmic processes, osmiophilic organelles (granules) and cisternal system. D = Granules; S = cistern; P = processes. From *Lerman et al.* (15).

kidneys), such as ours, a diligent search for RIC tumors gives a high yield of these structures (30% of 103 unselected cases had 'medullary fibromas'). In a thorough study, *Stuart et al.* (41) did not observe a relationship between the presence of the renomedullary tumor at autopsy and the existence of the hypertensive state during life.

Comment

Seven features of a schema considered necessary to establish a new endocrine system have been considered from a historical perspective. The renal medulla and its interstitital cells (RIC) appear to constitute an endocrine type antihypertensive organ by satisfying five of the seven features. The final proof (or the 'proof and counter proof' of *C. Bernard*) cannot come until the structural formula of the active principle of the RIC is established. Then it will be possible to demonstrate whether the antihypertensive substance (ANRL) extracted from the renal medulla and its RIC is, indeed, the mediator of the antihypertensive action of these cells. Since the action of ANRL mimics closely the action of the RIC, it could be that ANRL is the mediator of the antihypertensive action. As such, it remains the putative antihypertensive renomedullary hormone (ARH).

Summary

Ablation of renal tissue makes the subject sensitive to hypertension-inducing mechanisms, especially those due to fluid expansion, either by Na-volume or whole blood. Such hypertensive mechanisms are prevented by deviation of urine flow into a vein. Ablation of the renal medulla, by acute hydronephrosis or chemically, also potentiates the hypertensive state. Transplantation of either the renal medulla or its interstitial cells (RIC) can prevent or reverse hypertension. Under the latter conditions, the protective mechanism appears to result from the secretion of an antihypertensive hormone by the RIC. Lipid extracts of renal medulla not only prevent and reverse the hypertensive state in the same manner as medullary transplants but, under certain conditions, exert an acute depressor effect. The RIC can undergo hyperplastic changes much in the manner of an endocrine structure. For these reasons, it is proposed that the RIC represent an antihypertensive endocrine organ whose putative hormone may be termed the renomedullary antihypertensive hormone (ARH). Additional data in support of these contentions are presented.

References

1   Asimov, I.: Biographical encyclopedia of science and technology (Doubleday, Garden City 1964).
2   Berthold, A.A.: Transplantation der Hoden. Arch. Anat. Physiol. wiss. Med. 1849: 42 (1849).
3   Biedl, A.: The internal secretory organs: their physiology and pathology; translated by Linda Forster (William Wood, New York 1913).
4   Braun-Menendez, E. and Euler, U.S. von: Hypertension after bilateral nephrectomy in the rat. Nature, Lond. 160: 905 (1947).
5   Du Charme, D.W.: Personal commun.
6   Dunn, M.J.; Staley, R.S., and Harrison, M.: Characterization of prostaglandin production in tissue culture of rat renal medullary cells. Prostaglandins 12: 37–49 (1976).
7   Floyer, M.A.: Further studies on the mechanism of experimental hypertension in the rat. Clin. Sci. 14: 163–181 (1955).
8   Garrison, F.H.: History of medicine; 4th ed. (Saunders, Philadelphia 1929).
9   Grollman, A.; Muirhead, E.E., and Vanatta, J.: Role of the kidney in pathogenesis of hypertension as determined by a study of the effects of bilateral nephrectomy and other experimental procedures on the blood pressure of the dog. Am. J. Physiol. 157: 21–30 (1949).
10 Haggitt, R.C.; Pitcock, J.A., and Muirhead, E.E.: Renal medullary fibrosis in hypertension. Hum. Path. 2: 587–595 (1971).
11 Heptinstall, R.H.; Salyer, D.C., and Salyer, W.R.: The effects of chemical ablation of the renal papilla on blood pressure of rats with and without silver-clip hypertension. Am. J. Path. 78: 297–308 (1975).

12 *Kolff, W.J.:* The hypertension-reducing function of the kidney. Cleveland Clin. Q. *24:* 141–159 (1957).

13 *Kolff, W.J. and Page, I.H.:* Pathogenesis of renoprival cardiovascular disease in dogs. Am. J. Physiol. *178:* 75–81 (1954).

14 *Leonards, J.R. and Heisler, C.R.:* Maintenance of life in bilaterally nephrectomized dogs and its relation to malignant hypertension. Am. J. Physiol. *167:* 553–558 (1951).

15 *Lerman, R.I.; Pitcock, J.A.; Stephenson, P., and Muirhead, E.E.:* Renomedullary interstitial cell tumor. Hum. Path. *3:* 559–568 (1972).

16 *Maisel, A.Q.:* The hormone quest (Random House, New York 1965).

17 *Manger, W.M.; Praag, D.; Weiss, R.J.; Hart, C.J.; Hulse, C.I.; Roch, T.W., and Farber, S.J.:* Effect of transplanting renomedullary tissue into spontaneously hypertensive rats (SHR). Fed. Proc. Fed. Am. Socs exp. Biol. *35:* 556 (1976).

18 *Manthorpe, T.:* The effect on renal hypertension of subcutaneous isotransplantation of renal medulla on normal and hypertensive rats. Acta path. microbiol. scand., sect. A *81:* 725–733 (1973).

19 *Muehrcke, R.C.; Mandal, A.K., and Volini, F.I.:* A pathophysiological review of the renal medullary interstitial cells and their relationship to hypertension. Circulation Res. *26/27:* suppl. I, pp. 109–119 (1970).

20 *Muirhead, E.E.:* Protection against sodium-overload disease. Archs Path. *74:* 214–219 (1972).

21 *Muirhead, E.E.; Brooks, B.; Kosinski, M.; Daniels, E.G., and Hinman, J.W.:* Renomedullary antihypertensive principle in renal hypertension. J. Lab. clin. Med. *67:* 778–791 (1966).

22 *Muirhead, E.E.; Brooks, B.; Pitcock, J.A., and Stephenson, P.:* The renomedullary antihypertensive function in accelerated (malignant) hypertension. With observations on the renomedullary interstitial cells. J. clin. Invest. *51:* 181–190 (1972).

23 *Muirhead, E.E.; Brooks, B.; Pitcock, J.A.; Stephenson, P., and Brosius, W.L.:* Role of the renal medulla in the sodium-sensitive component of renoprival hypertension. Lab. Invest. *27:* 192–198 (1972).

24 *Muirhead, E.E.; Brown, G.B.; Germain, G.S., and Leach, B.E.:* The renal medulla as an antihypertensive organ. J. Lab. clin. Med. *76:* 641–651 (1970).

25 *Muirhead, E.E.; Daniels, E.G.; Pike, J.E., and Hinman, J.W.:* Renomedullary antihypertensive lipids and the prostaglandins; in *Bergstrom and Samuelsson* The prostaglandins. Nobel Symposium II (Almqvist & Wiksell, Stockholm 1967).

26 *Muirhead, E.E.; Germain, G.S.; Armstrong, F.B.; Brooks, B.; Leach, B.E.; Byers, L.W.; Pitcock, J.A., and Brown, P.:* Endocrine-type antihypertensive function of renomedullary interstitial cells. Kidney int. *8:* suppl. 5, pp. 272–282s (1975).

27 *Muirhead, E.E.:* The role of the renal medulla in hypertension; in *Stollerman* Advances in internal medicine (Year Book Med. Publ., Chicago 1974).

28 *Muirhead, E.E.; Hinman, J.W., and Daniels, E.G.:* Renoprival hypertension and the antihypertensive function of the kidney. Boerhaave Cursus: Hypertension. Boerhaave Kwart. Edit. (Leiden 1963).

29 *Muirhead, E.E.; Jones, F., and Stirman, J.A.:* Hypertensive cardiovascular disease of dog. Relation of sodium and dietary protein to uretero-caval anastomosis and ureteral ligation. Archs Path. *70:* 108–116 (1960).

30 *Muirhead, E.E.; Jones, F., and Stirman, J.A.:* Antihypertensive property in renoprival hypertension of extract from renal medulla. J. Lab. clin. Med. *56:* 167–180 (1960).

31 *Muirhead, E.E.; Leach, B.E.; Byers, L.W.; Brooks, B.; Daniels, E.G., and Hinman, J.W.:* Antihypertensive neutral renomedullary lipids (ANRL); in *Fisher* Kidney hormones (Academic Press, London 1970).

32 *Muirhead, E.E.; Leach, B.E.; Daniels, E.G., and Hinman, J.W.:* Lapine renomedullary lipid in murine hypertension. Archs Path. *85:* 72–79 (1968).

33 *Muirhead, E.E.; Rightsel, W.A.; Leach, B.E.; Byers, L.W.; Pitcock, J.A., and Brooks, B.:* The renomedullary antihypertensive function and its candidate antihypertensive hormone. Ann. Acad. Med. Singapore *5:* 36–44s (1976).

34 *Muirhead, E.E.; Rightsel, W.A.; Leach, B.E.; Byers, L.W.; Pitcock, J.A., and Brooks, B.:* Reversal of hypertension by transplants and lipid extracts of cultured renomedullary interstitial cells. Lab. Invest. *36:* 162–172 (1977).

35 *Muirhead, E.E.; Stirman, J.A., and Jones, F.:* Renal autoexplantation and protection against renoprival hypertensive cardiovascular disease and hemolysis. J. clin. Invest. *39:* 266–281 (1960).

36 *Muirhead, E.E.; Stirman, J.A.; Jones, F., and Lesch, W.:* The reduction of postnephrectomy hypertension by renal homotransplant. Surgery Gynec. Obstet. *103:* 673–686 (1956).

37 *Muirhead, E.E.; Vanatta, J., and Grollman, A.:* Papillary necrosis of the kidney. A clinical and experimental correlation. J. Am. med. Ass. *142:* 627–631 (1950).

38 *Pitcock, J.A.; Brown, P.; Brooks, B.; Brosius, W.L.; Clapp, W.L., and Muirhead, E.E.:* Partial nephrectomy-salt hypertension. Sodium balance and effect on renomedullary interstitial cells. Fed. Proc. Fed. Am. Socs exp. Biol. *36:* 438 (1977).

39 *Prezyna, A.; Attalah, A.; Vance, V.K.; Schoolman, M., and Lee, J.B.:* The renomedullary body. A newly recognized structure of renomedullary interstitial cell organ associated with high prostaglandin content. Prostaglandin *3:* 669–678 (1973).

40 *Solez, K.; D'Agostini, R.J.; Buono, R.A.; Vernon, N.; Wang, A.L.; Finer, P.M., and Heptinstall, R.H.:* The renal medulla and mechanisms of hypertension in the spontaneously hypertensive rat. Am. J. Path. *85:* 555–567 (1976).

41 *Stuart, R.; Salyer, W.R.; Salyer, D.C., and Heptinstall, R.H.:* Renomedullary interstitial cell lesions and hypertension. Hum. Path. *7:* 327–332 (1976).

42 *Susic, D.; Sparks, J.C., and Machado, E.A.:* Salt-induced hypertension in rats with hereditary hydronephrosis. The effect of renomedullary transplantation. J. Lab. clin. Med. *87:* 232–239 (1976).

43 *Tobian, L. and Azar, S.:* Antihypertensive and other functions of the renal papilla. Trans. Ass. Am. Physns *84:* 281–288 (1972).

E.E. Muirhead, MD, Professor of Pathology and Clinical Professor of Medicine, University of Tennessee, Center for the Health Sciences, 800 Madison Avenue, *Memphis, TN 38163* (USA)

Contr. Nephrol., vol. 12, pp. 82–91 (Karger, Basel 1978)

The Renal Prostaglandin System: Localization and Some Biological Effects[1]

Carin Larsson

Department of Pharmacology, Karolinska Institutet, Stockholm

Introduction

The discovery of renal prostaglandin (PG)-like compounds was reported in 1965 by *Lee et al.* (1), who also later identified them to be PGA_2, PGE_2 and $PGF_{2\alpha}$. Since then their biologically active intermediates, the PG-endoperoxides PGG_2 and PGH_2 (2, 3) and their active products, thromboxanes (4) and prostacyclin (PGI_2) (5, 6) have been discovered in some other tissues. The amount and kind of products formed in different parts of the kidney depends on the activity of each specific enzyme involved in the PG system in each special region of the kidney. Because of possible differences in effects, depending upon the kind of product and the site of production, the knowledge of the locality for the different PGs or their intermediates is of importance.

The renal PG system is one of several hormonal systems which locally in the kidney has been proposed to regulate renal blood flow and the homeostasis of body sodium and body water (7). To study the importance of the PGs in the renal functions, the activity of the endogenous PG system can be either enhanced by the endogenous precursor arachidonic acid (C20:4) (8) or decreased by PG biosynthesis inhibitors like indomethacin (9, 10).

C20:4 stimulates the endogenous renal PG biosynthesis at the sites where the necessary biosynthetic apparatus is available. Hereby is stimulated the synthesis of not only the well-know 'end-products' PGE_2, $PGF_{2\alpha}$ and PGD_2, but also possibly PGI_2 and thromboxanes and their common intermediates PGG_2 and PGH_2.

[1] These studies were supported by the Swedish Medical Research Council grant No. 2211.

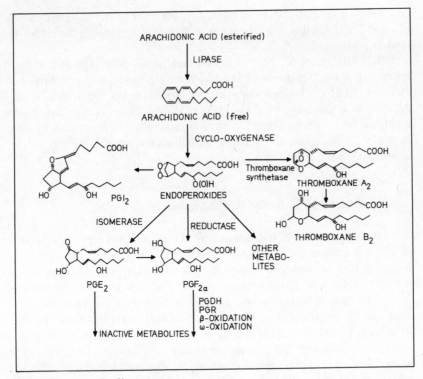

Fig. 1. The prostaglandin system.

The Prostaglandin System

The substrates, products and enzymes necessary for the formation and metabolism of the PGs can be referred to as the PG system. The precursor for the renal PG system is C20:4. Its main components are indicated in the scheme shown in figure 1. It was shown that the conversion to PGs occurred only with the free acid following its release from complex lipids, e.g. phospholipids (11, 12). The rate-limiting factor in the formation of PGs appears to be the availability of the precursor acid at the site of formation (13). A cyclo-oxygenase converts C20:4 to an intermediate, the PG-endoperoxide PGG_2 (2, 3). PGE_2 and the isomer, PGD_2, is formed by the reduction of the 9−11 peroxygroup to a keto- and a hydroxygroup. $PGF_{2\alpha}$ is formed by reduction of the peroxygroup to two hydroxygroups. The formation of PGE_2 and $PGF_{2\alpha}$ can also be influenced by PGE-9-keto-reductase (14), which catalyzes the conversion of PGE_2 to $PGF_{2\alpha}$. In many organs PGG_2 can be converted to thromboxane A_2 (TXA_2) which is very rapidly converted to thromboxane B_2 (TXB_2) (4). PGG_2 may also be converted to PGI_2, which is rapidly transformed to the more stable 6-keto-$PGF_{1\alpha}$ (5, 6). Whether thromboxanes or PGI_2 play any role in the kidney is not clear at the present.

The metabolism of the PGs occurs rapidly following their formation. The initial step is the conversion of the 15-hydroxylgroup to a ketone, catalyzed by PGDH (15, 16). The next step in the metabolism is the reduction of the Δ^{13}-double bond, catalyzed by PGR (15, 16). These conversions represent biological inactivation of the PGs. Further metabolism of PGs occurs by β-oxidation and ω-oxidation to form corresponding dicarboxyl PGs with 16 carbon atoms, the main metabolites in urine.

Subcellular Distribution of the PG System

The PG precursor C20:4 is an essential fatty acid and represents about 15% of all fatty acids in the kidney, a proportion which is higher than in most other organs. C20:4 is mainly esterified to phospholipids (17). Only in the 'lipid droplets' which have lower amounts of C20:4, C20:4 occurs in the triglyceride form. The lipid droplets do not produce or store PGs, but they may serve to store the PG precursor (17). The complex of enzymes which form the PGs are membrane bound and found mainly in the endoplasmatic reticulum (17, 18). After the biosynthesis the PGs are not stored but at once set free to the cytosol, where they are quickly metabolized by locally available PGDH. The site of action of the PGs is therefore mostly in direct connection with the site of synthesis.

Regional Distribution of the PG System

PG formation, measured by bioassay, is highest in the papilla with slightly lower levels in the outer medulla (19). The collecting tubules (20) and the interstitial cells (21) have been proposed as main cellular localization for PG biosynthesis.

Only small amounts of smooth muscle activity were found in the cortex (19). However, when PG formation was measured as the conversion by microsomes of tritium-labelled C20:4 to PGs and PG metabolites, the activity in the cortex was one third of that in the medulla (19) and thus about as high as PG formation in the lung. The high levels of PGDH in the cortex (19, 22) — only low activity is found in the medulla — results in rapid metabolism. Since in the radiochemical assay metabolites are also included in the product, the discrepancy between the two methods is most likely explained by the rapid metabolism of PGs in the renal cortex.

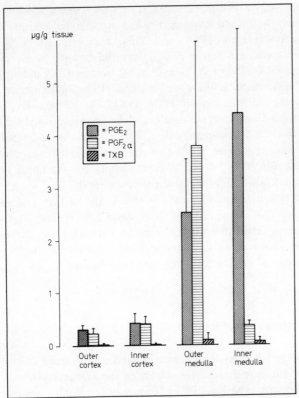

Fig. 2. Distribution of PGE_2, $PGF_{2\alpha}$ and TXB_2 in four different regions of rabbit kidney.

The endogenous amounts of PGs in cortex and medulla from rabbit kidney were measured by mass fragmentography and the PGs identified as PGE_2 and $PGF_{2\alpha}$ (23). Following amounts ($\mu g/g$) were found: in cortex 0.19 ± 0.04 PGE_2 and 0.21 ± 0.007 $PGF_{2\alpha}$, and in medulla 4.36 ± 1.04 PGE_2 and 1.64 ± 0.50 $PGF_{2\alpha}$. The amount of PGA_2 measured could be accounted for by the nonenzymatic conversion from PGE_2. This has also been confirmed by others (24, 25). PGD_2 was not assayed in this study, but others have demonstrated its presence also in the kidney (26, 27). The levels of PGs represent the net of synthesis and metabolism. Therefore, the detected values of PGE_2 and $PGF_{2\alpha}$ in the cortex, with very active PGDH, must be considered as a minimum of the true formation there. The presence of PG biosynthesis activity in the renal cortex provides the biochemical basis for an effect on the renal hemodynamics and interaction there with the renin-angiotensin system.

In another study TXB_2, the stable metabolite of TXA_2, was included in the measurement of PGs in outer and inner cortex and medulla, respectively (28) (fig. 2). In the cortex similar levels of PGE_2 and $PGF_{2\alpha}$ were detected. PGE_2 was highest in the inner medulla whereas $PGF_{2\alpha}$ was the highest in the outer medulla. This regional difference of PGE_2 and $PGF_{2\alpha}$ is interesting and may be the consequence of differences in the distribution of either the isomerase and the reductase, catalyzing the formation of PGE_2 and $PGF_{2\alpha}$, respectively, from C20:4, or of the PGE-9-keto-reductase catalyzing the conversion of PGE_2 to $PGF_{2\alpha}$. The activity of PGE-9-keto-reductase in the rabbit kidney is altered by changes in sodium balance (29) (see also *Weber,* this volume). It is tempting to speculate about a role for $PGF_{2\alpha}$ in, or in close vicinity to, the upper distal tubule, where sodium is known to influence the release of renin.

Almost no TXB_2 was detected. Incubations with C20:4 or PGH_2 increased PGE_2 and $PGF_{2\alpha}$, but TXB_2 remained low. *Needleman* (30) recently reported that in a hydronephrotic kidney the thromboxane formation is highly increased following bradykinin. However, no thromboxanes were detected in the normal kidney. Thus, the types of PGs formed can vary from normal to pathological conditions.

Some Biological Effects of the Renal PG System

There are several possibilities for pharmacological modification of the PG system. The activity can be enhanced by increasing the availability of the PG precursor, by administering exogenous PG-endoperoxides or PGs, or possibly by inhibiting PGDH. The differences in the PG system in the various parts of the kidney makes it difficult, with the proper exogenous PG, to reach their normal endogenous site of action. Also, the effects can be secondary to their hemodynamic activity. It seems therefore preferable to use endogenous precursors since then the biosynthesis of PG intermediates and their possible products is stimulated at the endogenous site of synthesis and in the direction specific for each particular region.

Another way to study the effects of endogenous PGs or their intermediates is to inhibit their biosynthesis (9, 10). There are different inhibitors for the various biosynthetic steps on the PG system. By using PG synthesis inhibitors as indomethacin or 5,8,11,14-eicosatetraynoic acid (ETA), the following biological roles for the PGs in the kidney have been postulated (see elsewhere in this symposium): (1) modulation of the action of the antidiuretic hormone; (2) modulation of the action of vasopressor hormones; (3) determinants of magnitude and distribution of renal blood flow (RBF), and (4) interaction with the renin-angiotensin system. I report the use of C20:4 to study the effects of endogenous renal PGs on RBF distribution and on renin release.

Table I. Effect of arachidonic acid (C20:4) and indomethacin in rabbit on PG efflux in urine, distribution of renal cortical blood flow (RBF) and plasma renin activity (PRA)

Treatment	Urinary PG, %	RBF, % inner cortex outer cortex	PRA, ng/ml/h
Control	107 ± 5.2	95.1 ± 3.1	45 ± 7.3
C20:4	150 ± 15	112 ± 2.3	91 ± 16
	(p < 0.05)	(p < 0.001)	(p < 0.05)
C20:4 + indomethacin	58 ± 12	78 ± 7.1	
Indomethacin			27 ± 6.0
			(p < 0.001)

Effect of Arachidonic Acid (C20:4) and Indomethacin on RBF

Arterial infusion of nonhypotensive doses of C20:4 (10–20 µg/kg/min), to anesthetized rabbits (31) and dogs (32), proximal to the renal arteries, increases the endogenous production of PGs in the kidney (table I) (31), detected as the urinary PG outflow (33). RBF, measured with radioactively labelled microspheres, is also increased and distributed to the juxtamedullary cortex (31, 32) (table I), possibly through decreased vascular resistance there or in the medulla. In contrast, pretreatment with indomethacin decreased urinary PG outflow and also RBF (31, 34), especially in the juxtamedullary region (31, 35–38). Stimulated renal endogenous PG production has also been noticed when RBF is decreased by stimuli, frequently of a supraphysiologic nature (7). Also anesthesia as such may induce PG formation, since indomethacin did not decrease RBF in conscious dogs (39). Thus, the biological significance for the renal PGs to exert their vasodilatory effects may be restricted to certain conditions with reduced RBF.

In vivo Effects of C20:4, PGH₂, Renal Clamping and Indomethacin on Renin

Nonhypotensive doses of C20:4 to anesthetized rabbits (40) or rats (41) increases plasma renin activity (PRA) (table I) (40). PRA was measured with radioimmunoassay as the amount of angiotensin I produced following incubation of plasma (42). In contrary, indomethacin decreases PRA.

In separate experiments nonhypotensive doses of the PG-endoperoxide PGH_2 (kindly supplied by Dr. Mats Hamberg), injected into the renal arteries, increased plasma renin concentration (PRC) dose-dependently in anesthetized,

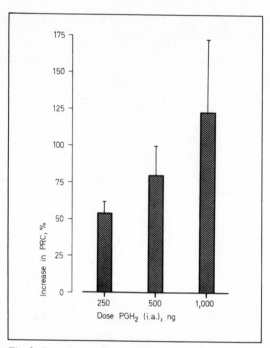

Fig. 3. Dose-dependent increase (%) in plasma renin concentration (PRC) following administration of PGH₂ into the renal artery of the rabbit.

indomethacin-treated rabbits (fig. 3): 54 ± 9% (250 ng PGH$_2$), 80 ± 21% (500 ng PGH$_2$) and 123 ± 49% (1,000 ng PGH$_2$). PRC was measured as PRA but with the difference that an excess of purified rabbit renin substrate (kindly supplied by Dr. *Weber*) was added to plasma when incubated (42).

Clamping of the renal artery, or renal ischemia, has also shown an increase in PRA (43) and PG release (44, 45). The effect of indomethacin and total renal clamping on PRA was studied in anesthetized rabbits (46). 20 min following release of the clamping of both renal pedicles (n = 6), PRA increased from 81 ± 14 to 121 ± 15 ng/ml/h (mean ± SEM; p <0.05), with a maximum after 60–90 min (46). Pretreatment with indomethacin (5 mg/kg; n = 7) decreased PRA from 121 ± 22 to 59 ± 11 ng/ml/h. Mass fragmentographic analysis of PGE$_2$ content in the kidneys indicated that indomethacin had inhibited the PG biosynthesis by at least 90% (46).

The renal perfusion presssure may be decreased by reduced vascular resistance, as by C20:4, or by clamping the renal artery, interventions which were shown to stimulate the renal endogenous PG production. Reduction in renal perfusion pressure and distension of the renal arterioles is also known to increase renin release. This mechanism for increase in renin release may involve the PGs and be either secondary to their hemodynamical effect or be caused by a direct influence of the PGs on renin release. The possibility for the renal PG system to influence renin release directly was further studied and will be discussed by Dr. *Weber* (see pp. 92—105).

Summary

The renal prostaglandins (PGs) are formed mainly in the endoplasmatic reticulum from locally available precursor, arachidonic acid (C20:4). Although the main PG formation occurs in the papilla, significant amounts of PGs are also formed in the cortex. PGs are not stored, but at once released to the cytosol and metabolized by soluble enzymes, 15-OH-PG-dehydrogenase (PGDH), Δ^{13}-PG-reductase and PGE-9-keto-reductase. PG metabolism by the PGDH pathway occurs predominantly in cortex.

C20:4 can be used to study the biological effects of the renal PG system. C20:4 given to rabbits increases renal biosynthesis of PGs, renal blood flow, predominantly in the juxtamedullary cortex, and plasma renin activity. These effects are inhibited by PG synthesis inhibitors like indomethacin.

References

1   *Lee, J.B.; Covino, B.G.; Takman, B.H., and Smith, F.R.:* Renomedullary vasodepressor substance, medullin. Isolation, chemical characterization and physiological properties. Circulation Res. *17:* 57—77 (1965).

2   *Hamberg, M. and Samuelsson, B.:* Detection and isolation of an endoperoxide intermediate in prostaglandin biosynthesis. Proc. natn. Acad. Sci. USA *70:* 899—903 (1973).

3   *Nugteren, D.H. and Hazelhof, E.:* Isolation and properties of intermediates in prostaglandin biosynthesis. Biochim. biophys. Acta *326:* 448—461 (1973).

4   *Hamberg, M.; Svensson, J., and Samuelsson, B.:* Thromboxanes; a new group of biologically active compounds derived from prostaglandin endoperoxides. Proc. natn. Acad. Sci. USA *72:* 2994—2998 (1975).

5   *Pace-Asciak, C. and Wolfe, L.S.:* A novel prostaglandin derivate formed from arachidonic acid by rat stomach homogenates. Biochemistry, N.Y. *10:* 3657—3664 (1971).

6   *Moncada, S.; Gryglewski, R.; Bunding, S., and Vane, J.R.:* An enzyme isolated from arteries transforms prostaglandin endoperoxides to an unstable substance that inhibits platelet aggregation. Nature, Lond. *263:* 663—665 (1976).

7   *Anderson, R.J.; Berl, T.; McDonald, K.M., and Schrier, R.:* Prostaglandins: effects on blood pressure, renal blood flow, sodium and water excretion. Kidney int. *10:* 205—214 (1977).

8   *Larsson, C. and Änggård, E.:* Arachidonic acid lowers and indomethacin increases the blood pressure of the rabbit. J. Pharm. Pharmac. *25:* 653—655 (1973).

9   *Vane, J.R.:* Inhibition of prostaglandin synthesis as a mechanism of action for aspirin-like drugs. Nature new Biol. *231:* 232—235 (1971).

10 *Smith, J.B. and Willis, A.L.:* Aspirin selectively inhibits prostaglandin production in human platelets. Nature new Biol. *231:* 235—237 (1971).

11 *Lands, W.E.M. and Samuelsson, B.:* Phospholipid precursors of prostaglandins. Biochim. biophys. Acta *164:* 426–429 (1968).

12 *Vonkeman, H. and Dorp, D.A. van:* The action of prostaglandin synthetase on 2-arachidonyl-lecithin. Biochim. biophys. Acta *164:* 430–432 (1968).

13 *Samuelsson, B.:* Biosynthesis and metabolism of prostaglandins; in Proc. 4th Int. Cong. Pharmacol., Basel, pp. 12–31 (Schwabe, Basel 1970).

14 *Leslie, C.A. and Levine, L.:* Evidence for the presence of a prostaglandin E_2-9-keto-reductase in rat organs. Biochem. biophys. Res. Commun. *52:* 717–724 (1973).

15 *Änggård, E. and Larsson, C.:* The sequence of the early steps in the metabolism of prostaglandin E_1. Eur. J. Pharmacol. *14:* 66–70 (1971).

16 *Hamberg, M. and Samuelsson, B.:* Metabolism of prostaglandin E_2 in guinea pig liver. II. Pathways in the formation of the major metabolites. J. biol. Chem. *246:* 1073–1077 (1971).

17 *Änggård, E.; Bohman, S.-O.; Griffin, J.E., III; Larsson, C., and Maunsbach, A.B.:* Subcellular localization of the prostaglandin system in the rabbit renal papilla. Acta physiol. scand. *84:* 213–246 (1972).

18 *Bohman, S.-O. and Larsson, C.:* Prostaglandin synthesis in membrane fractions from the rabbit renal medulla. Acta physiol. scand. *94:* 244 (1975).

19 *Larsson, C. and Änggård, E.:* Regional differences in the formation and metabolism of prostaglandins in the rabbit kidney. Eur. J. Pharmacol. *21:* 30–36 (1973).

20 *Janzen, F.H.A. and Nugteren, D.H.:* Histochemical localisation of prostaglandin synthetase. Histochemistry *27:* 159–164 (1971).

21 *Muirhead, E.E.; Germain, G.; Leach, B.E.; Pitcock, J.A.; Stephenson, P.; Brooks, B.; Brosius, W.L.; Daniels, E.G., and Hinman, J.W.:* Production of renomedullary prostaglandins by renomedullary interstitial cells grown in tissue culture. Circulation Res. *31:* suppl. II, pp. 161–172 (1972).

22 *Änggård, E.; Larsson, C., and Samuelsson, B.:* The distribution of 15-hydroxyprostaglandin dehydrogenase and prostaglandin-Δ^{13}-reductase in tissues of the swine. Acta physiol. scand. *81:* 396–404 (1971).

23 *Larsson, C. and Änggård, E.:* Mass spectrometric determination of prostaglandin E_2, $F_{2\alpha}$ and A_2 in the cortex and medulla of the rabbit kidney. J. Pharm. Pharmac. *28:* 326–328 (1976).

24 *Frölich, J.C.; Sweetman, B.J.; Smigel, M.; Williams, W.M.; Carr, K., and Oates, J.A.:* Renal prostaglandin synthesis – analysis by competitive protein binding assay and gas chromatography-mass spectrometry. VIth Int. Conf. on Prostaglandins, Florence 1975, p. 19.

25 *Steffenrud, S.:* Method for gas chromatographic mass-spectrometric quantitation of PGA_2. VIth Int. Conf. on Prostaglandins, Florence 1975, p. 61.

26 *Blackwell, G.J.; Flower, R.J., and Vane, J.R.:* Some characteristics of the prostaglandin synthesizing system in rabbit kidney microsomes. Biochim. biophys. Acta *398:* 178 (1975).

27 *Tai, H.-H.; Tai, C.L., and Hollander, C.S.:* Biosynthesis of prostaglandins in rabbit kidney medulla. Biochem. J. *154:* 257 (1976).

28 *Larsson, C.; Änggård, E.; Hamberg, M., and Samuelsson, B.:* Unpublished observations.

29 *Weber, P.; Larsson, C., and Scherer, B.:* Prostaglandin E_2-9-ketoreductase as a mediator of salt intake – related prostaglandin-renin interaction. Nature, Lond. (in press, 1977).

30 *Needleman, P.:* Thromboxane synthesis by the kidney. 1977 Winter Prostaglandin Conference, Vail, Colo. 1977.

31 *Larsson, C. and Änggård, E.:* Increased juxtamedullary blood flow on stimulation of intrarenal prostaglandin biosynthesis. Eur. J. Pharmacol. *25:* 326–334 (1974).

32 *Chang, L.C.T.; Splawinski, J.A.; Oates, J.A., and Nies, A.S.:* Enhanced renal prostaglandin production in the dog. II. Effects on intrarenal hemodynamics. Circulation Res. *36:* 204–207 (1975).

33 *Frölich, J.C.; Sweetman, B.J.; Carr, K.; Splawinski, J.; Watson, J.T.; Änggård, E., and Oates, J.A.:* Occurrence of prostaglandins in human urine. Adv. Biosci. *9:* 321–330 (1973).

34 *Lonigro, A.J.; Itskovitz, H.D.; Crowshaw, K., and McGiff, J.C.:* Dependency of renal blood flow on prostaglandin synthesis in the dog. Circulation Res. *32:* 712–717 (1973).

35 *Itskovitz, H.D.; Terragno, N.A., and McGiff, J.C.:* Effect of a renal prostaglandin on distribution of blood flow in the isolated canine kidney. Circulation Res. *34:* 770–776 (1974).

36 *Itskovitz, H.D. and McGiff, J.C.:* Hormonal regulation of the renal circulation. Circulation Res. *34/35:* suppl. I, pp. 65–73 (1974).

37 *Kirschenbaum, M.A.; White, N.; Stein, J.H., and Ferris, T.F.:* Redistribution of renal cortical blood flow during inhibition of prostaglandin synthesis. Am. J. Physiol. *227:* 801–805 (1974).

38 *Solez, K.; Fox, J.A.; Miller, M., and Heptinstall, R.H.:* Effects of indomethacin on renal inner medullary plasma flow. Prostaglandins *7:* 91–98 (1974).

39 *Zins, G.R.:* Renal prostaglandins. Am. J. Med. *58:* 14–24 (1975).

40 *Larsson, C.; Weber, P., and Änggård, E.:* Arachidonic acid increases and indomethacin decreases plasma renin activity in the rabbit. Eur. J. Pharmacol. *28:* 391–394 (1974).

41 *Weber, P.; Holzgreve, H.; Stephen, R., and Herbst, R.:* Plasma renin activity and renal sodium and water excretion following infusion of arachidonic acid in rats. Eur. J. Pharmacol. *34:* 299–304 (1975).

42 *Weber, P.; Held, E.; Uhlich, E., and Eigler, J.O.C.:* Reaction constants of renin in juxtaglomerular apparatus and plasma renin activity after renal ischemia and hemorrhage. Kidney int. *7:* 331–341 (1975).

43 *Stein, J.H. and Ferris, T.F.:* The physiology of renin. Archs intern. Med. *131:* 860–872 (1973).

44 *McGiff, J.C.; Crowshaw, K.; Terragno, N.A.; Lonigro, A.J.; Strand, J.C.; Williamson, M.A.; Lee, J.B., and Ng, K.K.:* Prostaglandin-like substances appearing in canine renal venous blood during renal ischemia. Circulation Res. *27:* 765–782 (1970).

45 *Jaffe, B.M.; Parker, C.W.; Marshall, G.R., and Needleman, P.:* Renal concentrations of prostaglandin E in acute and chronic renal ischemia. Biochem. biophys. Res. Commun. *49:* 799–805 (1972).

46 *Larsson, C.; Weber, P., and Änggård, E.:* Stimulation and inhibition of renal PG biosynthesis. Effects on renal blood flow and on plasma renin activity. Acta biol. med. germ. *35:* 1195–1200 (1973).

Carin Larsson, PhD, Pharmacology Research Division, AB Kabi, *S–112 87 Stockholm* (Sweden)

Contr. Nephrol., vol. 12, pp. 92–105 (Karger, Basel 1978)

Renal Prostaglandins in the Control of Renin[1]

P.C. Weber

Department of Internal Medicine, University Hospital Innenstadt, University of
Munich, Munich

Introduction

Renin is produced in the juxtaglomerular apparatus (JGA), a structure
situated adjacent to the glomeruli, which are localized exclusively in the cortical
region of the kidney (5). The epithelioid cells, which regulate synthesis, storage
and release of renin in the JGA, are specialized smooth muscle cells in the
afferent (and efferent) arteriole of the glomerulus. The JGA also contains special
cells of the distal tubule, the macula densa cells and a third cell type, the
lacis cells which fill the space between afferent and efferent arteriole and the
macula densa cells. Through the lacis cells, the JGA has anatomical connections
to the mesangial cells within the glomerulus (2, 51). Although many signals have
been defined which regulate renin release via a variety of different receptors
(baroreceptor, macula densa receptor, β- and α-adrenergic receptors), the biochemical mechanisms which transform these different signals into an increase or
a decrease of renin secretion are unknown.

Prostaglandin (PG) synthesis has been demonstrated in a variety of vascular
beds, isolated vascular walls and in isolated smooth muscle cells (1, 32). In the
kidney, PG production is highest in the kidney medulla and papilla but there is
also appreciable PG synthesis in the kidney cortex (24, 41, 62). The demonstration of both PG synthesis in vascular smooth muscle cells and PG synthetase
activity in the kidney cortex plus the observation that the renin-producing cells
are of smooth muscle origin raise the possibility of a direct interaction between
the PG and the renin-angiotensin systems in the kidney cortex.

In this study, pharmacological, biochemical, physiological and clinical evidence is presented that renal PG synthesis may play a pivotal role in the
mechanisms regulating the release of renin.

[1] Supported by the Deutsche Forschungsgemeinschaft, grant WE 681/2.

Material and Methods

The methods and materials used in these studies are described in detail elsewhere (26, 45, 56–61). Briefly the following should be summarized.

Animal experiments. Conscious and anesthetized rabbits were used in the *in vivo* experiments. For the *in vitro* experiments slices of rabbit renal cortex were used.

Studies in man. Studies were performed in patients with Bartter's syndrome (n = 4), patients with essential hypertension (n = 37) and in normal subjects (n = 22).

Analyses. Plasma renin activity (PRA) was determined by radioimmunoassay of A I; renin release from cortical slices was measured using a highly purified homologous renin substrate. Urinary PGE_2 and $PGF_{2\alpha}$ were analyzed by radioimmunoassay after extraction and group separation on silicic acid columns, directly ($PGF_{2\alpha}$) or after conversion of PGE_2 to either PGB_2 or to $PGF_{2\alpha}$. PGE_2 was also measured by bioassay.

Materials. The PG precursor, C20:4, was used *in vivo* and *in vitro* as the sodium salt after prior demethylation. The primary prostaglandins PGE_2 and $PGF_{2\alpha}$, the natural PG-endoperoxide PGG_2 and two structurally different, synthetic, stable PG-endoperoxide analogs, EPA I and EPA II, were used *in vitro*. The PG synthesis inhibitor, indomethacin, was used *in vivo* and *in vitro*. Furosemide was used *in vivo*. Statistical evaluation of the results was performed using standard techniques (* p <0.05; ** p <0.001).

Results and Discussion

Stimulation and Inhibition of Renal PG Synthesis: Effects on Renin Release in vivo *and* in vitro

In vivo Studies

Arachidonic acid increases PRA in normal, anesthetized rabbits when it is infused in subhypotensive doses into the aorta proximal to the origin of the renal arteries (26). In contrast, infusion of the PG synthesis inhibitor, indomethacin, in subhypertensive doses reduces PRA (fig. 1). In similar studies, C20:4 has been shown to increase the release of renin in rats (57) and in dogs (4). That enhanced renal PG formation is the cause of the increase in renin release after C20:4 is suggested by the finding that urinary excretion of PGs increases after the infusion of C20:4 (25). Sodium and water excretion remained unchanged after C20:4 infusion in anesthetized, normally hydrated rabbits and rats (26, 57), decreased in anesthetized, volume expanded rats – due to a reduction in renal blood flow and glomerular filtration rate (57) – and increased in dogs (4, 48). Therefore, directed changes in renal handling of sodium and water after C20:4 seem not to be primarily involved in the changes of renin release which increased under all five experimental conditions. However, species differences, the state of sodium balance and anesthesia seem to influence the effects of stimulated or inhibited PG synthesis on renal handling of sodium and water (4, 22, 26, 35, 36, 43, 48, 57). In the anesthetized rabbit and dog, the renal hemodynamic alterations following C20:4 are characterized by a decrease in renal vascular resistance (7, 25) which may indicate that the increase in renin release after C20:4

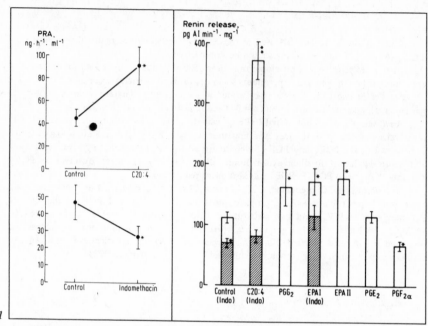

Fig. 1. Effects of infusion of arachidonic acid (C20:4) (upper panel) and of indomethacin (lower panel) on PRA in normal, anesthetized rabbits (mean ± SD).

Fig. 2. Effects of different compounds of the PG system on renin release from slices of rabbit kidney cortex without (whole columns) and with (shaded portion) pretreatment with indomethacin (mean ± SEM), explanation see text.

occurs secondary to changes in renal hemodynamics. Furthermore, it may be speculated that stimulated PG synthesis represents the biochemical mechanism for the increase of renin secretion under conditions which are characterized by decreased renal vascular resistance or by reduced renal perfusion pressure (10).

An increase of renal blood flow due to reduced renal vascular resistance is observed initially (10–15 min) after furosemide or similar diuretics (30, 36). The decrease in renal vascular resistance is paralleled by an increase of renin secretion (35, 36, 60). Both, the decrease in vascular resistance and the increase in renin release are most probably the result of stimulated renal PG production by these diuretics and can be blocked by PG synthesis inhibitors (30, 35, 36, 60).

These findings, together with observations that PG synthesis inhibitors reduce the renin secretion occurring after a variety of different stimuli such as total renal ischemia (27), sodium depletion (14), or isoproterenol infusion (13), suggest that PGs are important mediators in the release of renin in the case of each of the three known mechanisms for release of renin, baroreceptor, β-adrenergic and macula densa.

In vitro Studies

Stimulation of PG synthesis involves as the primary, rate-limiting step activation of a phospholipase which catalizes teh liberation of C20:4. Free C20:4, in turn, is converted by a cyclooxygenase into the PG-endoperoxide PGG_2, a biologically and chemically very unstable compound but biologically a very active intermediate in the PG system. PGG_2, in turn, is transformed by a variety of enzymes into the classical primary PGs PGE_2, PGD_2 and $PGF_{2\alpha}$ or into the PG-related compounds TXA_2 (18) and PGI_2 (32). The demonstration of each of these compounds in the kidney (3, 9, 24, 27, 33, 37, 47, 62) and their different effects in several biological systems makes conclusions from *in vivo* studies (using stimulation or inhibition of the renal PG system as a whole), concerning the specific function of the PGs in the renin release process, impossible.

In an attempt to examine the possibility that PGs are intrinsic to the process by which renin is released, and to characterize more specifically the possible biochemical interactions between the PG system and renin release, we studied the effects of stimulation and inhibition of PG synthesis and the effects of different compounds of the PG system on renin release from rabbit renal cortical slices *in vitro* (58).

Arachidonic acid (C20:4), the natural PG-endoperoxide PGG_2 and the synthetic, stable PG-endoperoxide analogs EPA I and EPA II stimulate renin release. PGE_2 has no effect and $PGF_{2\alpha}$ ($10^{-6} M$) inhibits renin release. The results for the maximally effective concentration of each compound on renin release are summarized in figure 2. C20:4 was the most potent of the compounds studied; it stimulated renin release up to a maximum of 350% as compared to control. It appears that roughly the same degree of stimulation was obtained with PGG_2, EPA I and EPA II (60%). In relation to the concentration used, PGG_2 and EPA I were about five times more active than EPA II. EPA I has been shown in other biological systems to mimic primarily the effects of PG-endoperoxides, whereas EPA II has been found to mimic primarily those of thromboxanes. Therefore, it seems that thromboxanes may not be the major compounds of the PG system to stimulate renin release. PGG_2 probably was rapidly degraded under these experimental conditions and the stimulation of this compound therefore was very short-lasting. In all experiments in which there was pretreatment with indomethacin, basal as well as C20:4 − or EPA I − stimulated renin release was reduced as compared to control. These results support the assumption that the increase of renin release after C20:4 is due to formation of PGs or PG-endoperoxides. Furthermore, it can be concluded that

EPA I owes part of its effects on renin release to stimulation of endogenous PG biosynthesis. Because of the significantly greater effect of C20:4 on renin release in comparison to the PG-endoperoxides, a role of other PG-related compounds such as prostacyclin (PGI_2) as a major effector of the PG system on renin release cannot be excluded.

In conclusion, these data raise the possibility that the renal cortical PG system is intrinsic to the mechanisms by which renin is released. The results further indicate that formation of PG-endoperoxides may stimulate, and formation of $PGF_{2\alpha}$ may inhibit renin secretion.

Role of Sodium Chloride Intake on Interaction between PGs and Renin

Changes in sodium and water balance are important determinants of the activity and function of the renin-angiotensin system (16, 51). The role of PGs in natriuresis and diuresis is a matter of controversy, and the question of a natriuretic or antinatriuretic action of PGs is at present unsolved (22, 28, 31, 39, 43, 49, 53).

To evaluate a possible functional relationship between the PG and the renin-angiotensin systems depending on the state of sodium balance, we determined the activities of both systems in rabbits kept on a low and a high sodium chloride intake (45, 59). The results are summarized in figure 3. The observed decrease in urinary PGE_2 excretion on high NaCl intake may result from either a decrease in renal PGE_2 synthesis, an alteration in PGE_2 metabolism at maintained synthesis, or both. The total amount of PGE_2 and $PGF_{2\alpha}$ excreted in urine decreased at high NaCl intake, suggesting a lower renal PG production. However, $PGF_{2\alpha}$ excretion remained unchanged at high NaCl intake. Together, these findings provide evidence of an altered PGE_2 metabolism on reduced PG synthesis at high NaCl intake. One factor influencing the pathway and thus, the consequences of altered PG formation, is the enzyme PGE_2-9-ketoreductase which catalyzes the interconversion of the antagonistically acting PGE_2 and $PGF_{2\alpha}$ (29). The increase of renal PGE_2-9-ketoreductase activity, as demonstrated in this study during high NaCl intake, shifts intrarenal PG formation from PGE_2 to $PGF_{2\alpha}$, and could explain the decrease in urinary PGE_2 excretion at unchanged $PGF_{2\alpha}$ excretion. The time course of activation of PGE_2-9-ketoreductase and PGE_2 excretion after increasing salt intake shows a delay with respect to changes in urinary excretion of sodium and may reflect enzyme-inductive processes necessary to increase PGE_2-9-ketoreductase activity. The relative increase of $PGF_{2\alpha}$ (ratio of PGE_2:$PGF_{2\alpha} \cong 1:1$ at low NaCl intake and 1:4 at high NaCl intake) on reduced PG synthesis may represent the biochemical process by which renin activity is reduced at positive sodium balance. At low NaCl intake the reverse may occur: PG synthesis increases, PGE_2-9-ketoreduc-

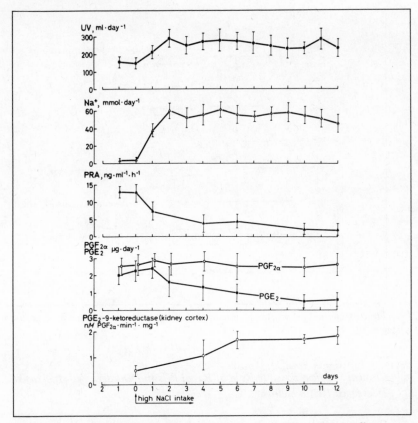

Fig. 3. Effects of dietary NaCl intake on urinary volume (UV), urinary sodium excretion (Na⁺), PRA and urinary PGE_2 and $PGF_{2\alpha}$ excretion and renal PGE_2-9-ketoreductase activity in normal rabbits (mean ± SD).

tase activity decreases, thereby increasing PGE_2 levels. Because of the stimulatory and relatively less opposed action of PG-endoperoxides on renin release, the stimulated renin activity could then initiate the sodium-conserving mechanism. A hypothetical scheme of the interactions between the PG and the renin-angiotensin systems, as well as the proposed role of sodium balance in the negative feedback control loop operating through aldosterone, extracellular volume and PGE_2-9-ketoreductase mechanisms, are depicted in figure 4. In conclusion we can state that the PG-endoperoxide stimulates and $PGF_{2\alpha}$ inhibits renin synthesis and/or release. The physiological meaning of these effects may depend upon an enzymatic, sodium-mediated control of conversion of PGE_2 to $PGF_{2\alpha}$.

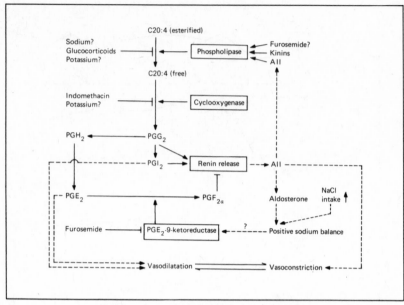

Fig. 4. Hypothetical scheme of interactions between the PG and the renin-angiotensin systems.

Clinical Studies: Possible Role of Renal PGs in Disorders with Increased or Decreased Renin Release

Bartter's Syndrome

This complex and rare disorder is characterized by renal losses of sodium, juxtaglomerular hyperplasia, hyperreninemia, secondary hyperaldosteronism with its consequences (hypokalemia, hyperkaliuria), resistance to the pressor action of angiotensin II and normal blood pressure. Increased renal PG synthesis appears primarily responsible for the major features of the syndrome: increased sodium excretion, excessive stimulation of renin release and decreased pressor responsiveness to angiotensin II (15, 55). Multiple reports have documented the beneficial effects of therapy with inhibitors of PG synthesis (12, 15, 17, 34, 55).

We studied the urinary excretion of PGE_2 and $PGF_{2\alpha}$ in 4 patients with Bartter's syndrome. In all patients excessive increases of renin and aldosterone, hypokalemia, and low blood pressure were present. In 2 patients in which angiotensin II was infused, the pressor response was significantly reduced. In 2 patients the diagnosis was verified by renal biopsy. The hyperplasia of the JGA including hypergranulation of the renin-producing cells in case 1 is shown in figure 5. In table I the values of urinary PG excretion in the 4 patients and in

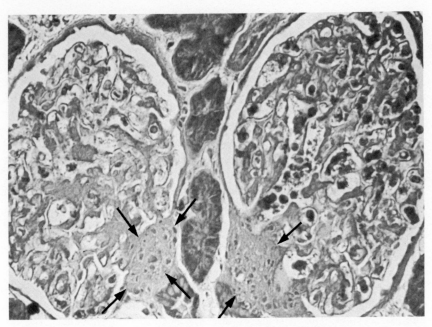

Fig. 5. Two glomeruli showing the hyperplasia of juxtaglomerular apparatuses in case 1 with Bartter's syndrome.

Table I. Urinary PG excretion in patients with Bartter's syndrome and in control subjects (mean ± SD)

	Case 1	Case 2	Case 3	Case 4	Control subjects (n = 20)
PGE_2, ng/24 h	1,733	426 ± 119 (n = 3 days)	780	290 ± 68 (n = 4 days)	280 ± 145
$PGF_{2\alpha}$, ng/24 h	810	152 ± 30	413	153 ± 25	510 ± 250
Ratio $PGE_2:PGF_{2\alpha}$	2.1:1	2.8:1	1.9:1	1.9:1	1:1.8

On indomethacin treatment (200 mg/day)

PGE_2
92 ± 40
(n = 4 days)

$PGF_{2\alpha}$
121 ± 37

Ratio

$PGE_2:PGF_{2\alpha}$
1:1.3

normal subjects are summarized. In all 4 patients with Bartter's syndrome an abnormal relation between PGE_2 and $PGF_{2\alpha}$ excretion was observed resulting in a mean ratio of $PGE_2:PGF_{2\alpha}$ of 2.2:1. In contrast, the mean ratio of $PGE_2:PGF_{2\alpha}$ was found to be 1:1.8 in normal persons. Absolute levels of urinary prostaglandins were elevated only in 2 patients (case 1, case 2). In patient (case 3), treated with indomethacin, the excretion of urinary PGs was reduced and the ratio of PGE_2 to $PGF_{2\alpha}$ shifted towards normal during therapy. At the same time PRA and urinary excretion of potassium and sodium decreased.

Thus, in Bartter's syndrome both, an increased formation of PG-endo-peroxides or PGE_2 as well as a relatively decreased formation of $PGF_{2\alpha}$ at elevated PG synthesis may represent primary pathogenetic mechanisms leading to hyperreninemia, renal sodium losses and diminished pressor responsiveness to angiotensin II. PG synthesis inhibitors probably corrects the major abnormalities, increased renal prostaglandin synthesis and hyperreninemia through its effects on prostaglandin metabolism.

Essential Hypertension

Subjects with essential hypertension can be classified into low, normal, or high renin subgroups based on the response of PRA to sodium restriction and upright posture. The meaning of this classification is under debate, and the resolution of such conventional techniques may be inadequate to detect pathogenetic mechanisms in this complex disease (6, 20, 23, 38, 52).

Recent reports have indicated that in the small number of patients with mild high renin 'essential' hypertension, increased renin release reflects autonomic overactivity which may be the primary pathogenetic factor in this form of hypertension (11, 42). On the other hand, it has been demonstrated that in the majority of patients with benign uncomplicated essential hypertension PRA is either normal or suppressed (21, 40, 44, 50). Furthermore, by using maneuvers which normally either increase or decrease renin release, a general unresponsiveness of renin secretion has been documented in these patients (21, 40, 54).

The notion of intrinsically produced renal vasoconstriction in essential hypertension seems consistent with the decreased sensitivity to renin-provoking stimuli (8, 19, 44). Furthermore, it may be speculated that the increase of vascular resistance in the kidney cortex as well as the decrease of renin release may be the result of the same pathophysiological defect.

The demonstration of a primary role of renal PGs in the release of renin and the finding that after furosemide or similar diuretics the initial increase of PRA which parallels the decrease in renal vascular resistance is the result of stimulation of renal PG synthesis (58, 60) formed the basis for our studies in patients with essential hypertension.

Before and after application of 40 mg furosemide i.v. we measured PRA in short time intervals in patients with mild uncomplicated essential hypertension

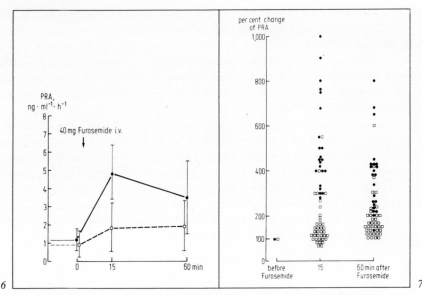

Fig. 6. Mean values (± SD) of absolute PRA levels in control subjects (•) and in patients with essential hypertension (□) before, 15 and 60 min after 40 mg furosemide i.v. (the subjects remained in the supine position throughout this time). Explanation see text.

Fig. 7. Scattergram of percent changes of PRA in control subjects (•) and in patients with essential hypertension (□), 15 and 60 min after 40 mg furosemide i.v.

(n = 37; age 22–57 years; 20 ♀, 17 ♂) and compared the results with those obtained in control subjects (n = 22; age 21–56 years; 12 ♀, 10 ♂). The data are summarized in figures 6 and 7.

At comparable mean sodium excretion in 24-hour urine (control subjects 164 mM/day; essential hypertensive subjects 162 mM/day) mean PRA values obtained in essential hypertensive patients (0.98 ± 0.95 ng AI·ml^{-1}·h^{-1}) did not differ from those obtained in control subjects (1.1 ± 0.7). In contrast to the similar PRA values in the resting condition, a highly significant difference in the increase of PRA was found initially (15 min) after furosemide. Whereas PRA increased to 4.8 ± 1.4 in normal subjects it increased only to 1.8 ± 1.3 in patients with essential hypertension (fig. 6). Furthermore, PRA values decreased after 60 min in control subjects, but showed no decrease in patients with essential hypertension.

We conclude that the reduced increase of renin release observed 15 min after furosemide in 32 out of 37 patients with essential hypertension is the result of an impaired PG synthesis at the site or near the JGA in these patients. The defect in PG synthesis may include (1) a reduced local pool of the PG precursor fatty acid C20:4, (2) a reduced activity of a phospholipase responsible for the liberation of C20:4, (3) a reduced formation of PG-endoperoxides or other PG-related compounds with renin-stimulating activity and (4) a relative increase in $PGF_{2\alpha}$ formation. All these hypothetical alterations in PG synthesis or PG metabolism are capable to diminish renin secretion at a given stimulus. Furthermore, it can be assumed that such alterations in PG formation, if occurring at vascular sites in the kidney cortex, may increase renal vascular resistance and decrease renal blood flow.

In conclusion, these results raise the possibility that a reduced PG-endoperoxide synthesis or a relative increase of $PGF_{2\alpha}$ formation in the kidney cortex are likely mechanisms leading to both a reduced secretion of renin and an increase of renal vascular resistance in patients with essential hypertension. The beneficial effects of sodium depletion or diuretics in essential hypertensive patients with low or unresponsive PRA values may depend on their potency to increase PG synthesis or to decrease $PGF_{2\alpha}$ formation, either directly or secondary to the decrease in extracellular volume (46, 59).

Summary

The biochemical processes which transform baroreceptor, β-adrenergic and macula densa signals into an increase or a decrease of renin secretion are unknown. Evidence is presented that the renal PG system is intimately involved in the mechanisms regulating the release of renin. *In vivo* stimulation of renal PG synthesis by arachidonic acid (C20:4) or furosemide increases renin release. PG synthesis inhibitors decrease basal renin release and reduce the renin release following stimulation with C20:4, furosemide and renal ischemia. *In vitro*, C20:4 and the PG-endoperoxides stimulate renin release from the rabbit kidney cortex whereas $PGF_{2\alpha}$ inhibits it. This suggests an intrinsic role in the renin release mechanism of PGs, synthesized at or near the juxtaglomerular apparatus. The operation of this PG effect on renin release may depend upon a salt intake related control of PG synthesis and of conversion of PGE_2 to $PGF_{2\alpha}$. Increased or decreased renal PG synthesis may also be the primary event leading to elevated or reduced renin levels in some clinical disorders. In Bartter's syndrome, the elevated renin levels may result from an increase in PG synthesis or a decrease of $PGF_{2\alpha}$ formation. In benign, uncomplicated essential hypertension, decreased renal PG synthesis or increased $PGF_{2\alpha}$ formation may be the primary mechanism which reduces renin release and renal blood flow.

References

1 *Alexander, R.W. and Gimbrone, M.A., jr.:* Stimulation of prostaglandin E synthesis in cultured human umbilical vein smooth muscle cells. Proc. natn. Acad. Sci. USA *73:* 1617–1620 (1976).

2 *Barajas, L. and Latta, H.:* A three-dimensional study of the juxtaglomerular apparatus in the rat. Lab. Invest. *12:* 257–269 (1963).

3 *Blackwell, G.J.; Flower, R.J., and Vane, J.R.:* Some characteristics of the prostaglandin synthesizing system in rabbit kidney microsomes. Biochim. biophys. Acta *398:* 178–190 (1975).

4 *Bolger, P.M.; Eisner, G.M.; Ramwell, P.W., and Slotkoff, L.M.:* Effect of prostglandin synthesis on renal function and renin in the dog. Nature, Lond. *259:* 244–245 (1976).

5 Brown, J.J.; Davies, D.L.; Lever, A.F.; Parker, R.A., and Robertson, J.I.S.: The assay of renin in single glomeruli in the normal rabbit and the appearance of the juxtaglomerular apparatus. J. Physiol., Lond. 176: 418–428 (1965).

6 Brown, J.J.; Lever, A.F.; Robertson, J.I.S., and Schalekamp, M.A.: Pathogenesis of essential hypertension. Lancet 1217–1219 (1976).

7 Chang, L.C.T.; Splawinski, J.A.; Oates, J.A., and Nies, A.S.: Enhanced renal prostaglandin production in the dog. II. Effects on intrarenal hemodynamics. Cirulation Res. 36: 204–207 (1975).

8 Coleman, T.G.; Guyton, A.C.; Young, D.B.; Clue, J.W. de; Norman, R.A., jr., and Manning, R.D., jr.: The role of the kidney in essential hypertension. Clin. exp. Pharmacol. Physiol. 2: 571–581 (1975).

9 Dunn, M.J.: Renal prostaglandin synthesis in the spontaneously hypertensive rat. J. clin. Invest. 58: 862–870 (1976).

10 Eide, I.; Løyning, E., and Kiil, F.: Evidence for hemodynamic autoregulation of renin release. Circulation Res. 32: 237–245 (1973).

11 Esler, M.; Julius, S.; Zweifler, A.; Randall, O.; Harburg, E.; Gardiner, H., and Quattro, V. de: Mild high-renin essential hypertension. Neurogenic human hypertension? New Engl. J. Med. 296: 405–411 (1977).

12 Fichman, M.P.; Telfer, N.; Zia, P.; Speckart, P.; Golub, M., and Rude, R.: Role of prostaglandins in the pathogenesis of Bartter's syndrome. Am. J. Med. 60: 785–797 (1976).

13 Frölich, J.C.; Hollifield, J.W., and Oates, J.A.: Effect of indomethacin on isoproterol induced renin release. Clin. Res. 24: 9A (1976).

14 Frölich, J.C.; Hollifield, J.W.; Wilson, J.P.; Sweetman, B.J.; Seyberth, H.W., and Oates, J.A.: Suppression of plasma renin activity in man by indomethacin. Independence of sodium retention. Clin. Res. 24: 271A (1976).

15 Gill, J.R.; Frölich, J.C.; Bowden, R.E.; Taylor, A.A.; Keiser, H.R.; Seyberth, H.W.; Oates, J.A., and Bartter, F.C.: Bartter's syndrome. A disorder characterized by high urinary prostaglandins and a dependence of hyperreninemia on prostaglandin synthesis. Am. J. Med. 61: 43–51 (1976).

16 Gross, F.: Renin stores in the kidney and plasma renin activity; in Fisher Kidney hormones, pp. 102–116 (Academic Press, New York 1971).

17 Halushka, P.; Wholtman, H.; Margolius, H., and Privitera, P.: Increased urinary excretion of prostaglandin E-like material and kallikrein in Bartter's syndrome – effective treatment with indomethacin. Pharmacologist 18: 163A (1976).

18 Hamberg, M.; Svensson, J., and Samuelsson, B.: Thromboxanes, a new group of biologically active compounds derived from prostaglandin endoperoxides. Proc. natn. Acad. Sci. USA 72: 2994–2998 (1975).

19 Hollenberg, N.K. and Adams, D.F.: The renal circulation in hypertensive disease. Am. J. Med. 60: 773–784 (1976).

20 Jose, A.; Crout, J.R., and Kaplan, N.M.: Suppressed plasma renin activity in essential hypertension. Roles of plasma volume, blood pressure, and sympathetic nervous system. Ann. intern. Med. 72: 9–16 (1970).

21 Khokhar, A.M.; Slater, J.D.H.; Jowett, T.P., and Payne, N.N.: Suppression of the renin-aldosterone system in mild essential hypertension. Clin. Sci. molec. Med. 50: 269–276 (1976).

22 Kirschenbaum, M.A. and Stein, J.H.: The effect of inhibition of prostaglandin synthesis on urinary sodium excretion in the conscious dog. J. clin. Invest. 57: 517–521 (1976).

23 Laragh, J.H.: Modern system for treating high blood pressure based on renin profiling and vasoconstriction-volume analysis. A primary role for beta blocking drugs such as propranolol. Am. J. Med. 61: 797–810 (1976).

24   *Larsson, C. and Änggard, E.:* Regional differences in the formation and metabolism of prostaglandins in the rabbit kidney. Eur. J. Pharmacol. *21:* 30–36 (1973).

25   *Larsson, C. and Änggard, E.:* Increased juxtamedullary blood flow on stimulation of intrarenal prostaglandin biosynthesis. Eur. J. Pharmacol. *25:* 326–334 (1974).

26   *Larsson, C.; Weber, P.C., and Änggard, E.:* Arachidonic acid increases and indomethacin decreases plasma renin activity in the rabbit. Eur. J. Pharmacol. *28:* 391–394 (1974).

27   *Larsson, C.; Weber, P.C., and Änggard, E.:* Stimulation and inhibition of renal PG biosynthesis. Effects on renal blood flow and on plasma renin activity. Acta biol. med. germ. *35:* 1195–1200 (1976).

28   *Lee, J.B.; Patak, R.V., and Mookerje, B.K.:* Renal prostaglandins and the regulation of blood pressure and sodium and water homeostasis. Am. J. Med. *60:* 798–816 (1976).

29   *Lee, S.-C. and Levine, L.:* Prostaglandin metabolism. J. biol. Chem. *249:* 1369–1375 (1974).

30   *Ludens, J.H.; Heitz, D.C.; Brody, M.J., and Williamson, H.E.:* Differential effect of furosemide on renal and limb blood flows in the conscious dog. J. Pharmac. exp. Ther. *171:* 300–306 (1970).

31   *McGiff, J.C.; Crowshaw, K., and Itskowitz, H.D.:* Prostaglandins and renal function. Fed. Proc. Fed. Am. Socs exp. Biol. *33:* 39–47 (1974).

32   *Moncada, S.; Gryglewski, R.; Bunting, S., and Vane, J.R.:* An enzyme isolated from arteries transforms prostaglandin endoperoxides to an unstable substance that inhibits platelet aggregation. Nature, Lond. *263:* 663–665 (1976).

33   *Needleman, P.:* Thromboxane synthesis by the kidney. Annual Winter Conf. on Prostaglandins, Vail, Colo. (1977).

34   *Norby, L.; Lentz, R.; Flamenbaum, W., and Ramwell, P.:* Prostaglandins and aspirin therapy in Bartter's syndrome. Lancet *ii:* 604–606 (1976).

35   *Oliw, E.; Köver, G.; Larsson, C., and Änggard, E.:* Reduction by indomethacin of furosemide effects in the rabbit. Eur. J. Pharmacol. *38:* 95–100 (1976).

36   *Olsen, U.B. and Ahnfelt-Rønne, I.:* Bumetanide induced increase of renal blood flow in conscious dogs and its relation to local renal hormones (PGE, kallikrein and renin). Acta pharmac. tox. *38:* 219–228 (1976).

37   *Pace-Asciak, C.R.:* The formation of 6-OXO-PGF$_{1\alpha}$ in different tissues. Annual Winter Conf. on Prostaglandins, Vail, Colo. (1977).

38   *Padfield, P.L.; Beevers, D.G.; Brown, J.J.; Davies, D.L.; Lever, A.F.; Robertson, J.I.S.; Schalekamp, M.A.D., and Tree, M.:* Is low-renin hypertension a stage in the development of essential hypertension or a diagnostic entity? Lancet 548–550 (1975).

39   *Papanicolaou, N.; Lefkos, N.; Safar, M.; Paris, M.; Bariety, J., and Milliez, M.:* Direct relationship between urinary prostaglandin E and sodium excretion in essential hypertensive patients. Experientia *32:* 1280–1281 (1976).

40   *Pedersen, E.B. and Kornerup, H.J.:* Renal hemodynamics and plasma renin in patients with essential hypertension. Clin. Sci. molec. Med. *50:* 409–414 (1976).

41   *Pong, S.S. and Levine, L.:* Biosynthesis of prostaglandins in rabbit renal cortex. Res. Commun. chem. Pathol. Pharmacol. *13:* 115–123 (1975).

42   *Quattro, V. de; Campese, V.; Miura, Y., and Meijer, D.:* Increased plasma catecholamines in high renin hypertension. Am. J. Cardiol. *38:* 801–804 (1976).

43   *Rosenthal, J.; Simone, P.G., and Silbergleit, A.:* Effects of prostaglandin deficiency on natriuresis, diuresis and blood pressure. Prostaglandins *5:* 435–440 (1974).

44   *Schalekamp, M.A.D.H.; Schalekamp-Kuyken, M.P.A., and Birkenhäger, W.H.:* Abnormal renal haemodynamics and renin suppression in hypertensive patients. Clin. Sci. *38:* 101–110 (1970).

45 *Scherer, B.; Siess, W., and Weber, P.C.:* Radioimmunological and biological measurement of prostaglandins in rabbit urine. Decrease of PGE_2 excretion at high NaCl intake. Prostaglandins *13:* 1127–1139 (1977).

46 *Stone, K.H. and Hart, M.:* Inhibition of renal PGE_2-9-ketoreductase by diuretics. Prostaglandins *12:* 197–207 (1976).

47 *Tai, H.-H.:* Mechanism of prostaglandin biosynthesis in rabbit kidney medulla. Biochem. J. *160:* 577–581 (1976).

48 *Tannenbaum, J.; Splawinski, J.A.; Oates, J.A., and Nies, A.S.:* Enhanced renal prostaglandin production in the dog. I. Effects on renal function, Circulation Res. *36:* 197–203 (1975).

49 *Terashima, R.; Anderson, F.L., and Jubiz, W.:* Prostaglandin E release in the dog. Effect of sodium. Am. J. Physiol. *231:* 1429–1432 (1976).

50 *Thomas, G.W.; Ledingham, J.G.G.; Beilin, L.J., and Stott, A.N.:* Reduced plasma renin activity in essential hypertension – effects of blood pressure, age and sodium. Clin. Sci. molec. Med. *51:* suppl. 3; pp. 185–189 (1976).

51 *Thurau, K. and Mason, J.:* The intrarenal function of the juxtaglomerular apparatus; in *Thurau* International review of physiology, pp. 357–389 (Med. Techn. Pub., London 1974).

52 *Thurston, H. and Swales, J.D.:* Low renin hypertension – a distinct entity. Lancet 930–932 (1976).

53 *Tobian, L. and O'Donnell, M.:* Renal prostaglandins in relation to sodium regulation and hypertension. Fed. Proc. Fed. Am. Socs exp. Biol. *35:* 2388–2392 (1976).

54 *Tuck, M.L.; Williams, G.H.; Dluhy, R.G.; Greenfield, M., and Moore, T.J.:* A delayed suppression of the renin-aldosterone axis following saline infusion in human hypertension. Circulation Res. *39:* 711–717 (1976).

55 *Verberckmoes, R.; Damme, B. van; Clement, J.; Amery, A., and Michielsen, P.:* Bartter's syndrome with hyperplasia of renomedullary cells. Successful treatment with indomethacin. Kidney int. *9:* 302–307 (1976).

56 *Weber, P.C.; Bauer, J.; Uhlich, E.; Held, E., and Schildberg, F.:* Kidney and plasma renin in human renovascular hypertension. Eur. J. clin. Invest. *6:* 415–424 (1976).

57 *Weber, P.C.; Holzgreve, H.; Stephan, R., and Herbst, R.:* Plasma renin activity and sodium and water excretion following infusion of arachidonic acid in rats. Eur. J. Pharmacol. *34:* 299–304 (1975).

58 *Weber, P.C.; Larsson, C.; Änggard, E.; Hamberg, M.; Corey, E.J.; Nicolaou, K.C., and Samuelsson, B.:* Stimulation of renin release from rabbit renal cortex by arachidonic acid and prostaglandin endoperoxides. Circulation Res. *39:* 868–874 (1976).

59 *Weber, P.C.; Larsson, C., and Scherer, B.:* Prostaglandin E_2-9-ketoreductase as a mediator of salt intake-related prostaglandin-renin interaction. Nature, Lond. *266:* 65–66 (1977).

60 *Weber, P.C.; Scherer, B., and Larsson, C.:* Increase of free arachidonic acid by furosemide in man as the cause of prostaglandin and renin release. Eur. J. Pharmacol. *41:* 329–332 (1977).

61 *Weber, P.C.; Held, E.; Uhlich, E., and Eigler, J.O.C.:* Reaction constants of renin in juxtaglomerular apparatus and plasma renin activity after renal ischemia and hemorrhage. Kidney int. *7:* 331–341 (1975).

62 *Whorton, R.; Frölich, J.C., and Oates, J.A.:* Prostacyclin is produced in renal cortical microsomes. Annual Winter Conf. on Prostaglandins, Vail, Colo. (1977).

P.C. Weber, MD, Department of Internal Medicine, University Hospital Innenstadt, University of Munich, Ziemssenstrasse 1, *D–8000 Munich* (FRG)

Contr. Nephrol., vol. 12, pp. 106–115 (Karger, Basel 1978)

Cellular Interactions between Vasopressin and Prostaglandins in the Mammalian Kidney

Thomas P. Dousa and Thomas E. Northrup

Nephrology Research Laboratories, Department of Physiology and Biophysics and Division of Nephrology, Department of Internal Medicine, Mayo Clinic and Mayo Foundation, Rochester, Minn.

Introduction

Interactions between vasopressin (VP) and prostaglandins (PG)[1] in the kidney represent a classical example of interrelationships between the effect of a typical extrarenal hormone (VP), which regulates one of the major renal functions – water handling – and humoral agents (PG), which are generated within the kidney itself. Interest in the interactions between VP and PG was stirred by discoveries which delineated that antidiuretic effect of VP is mediated by generation of cyclic adenosine 3,5-monophosphate (cAMP) in the renal medulla (1), by the finding that the renal medulla, namely inner medulla (papilla), is one of the most active sites of PG synthesis (2, 52, 53, 54) and by observations that PG exert a modulatory effect in other hormonal systems mediated by cAMP at the cellular level (3).

We will discuss briefly our current knowledge about VP-PG interactions at the level of kidney function, in terms of control of water permeability of nephron (antidiuretic or hydro-osmotic effects) and then, specifically, mode of these interactions at the level of individual cellular components.

It is generally acknowledged that the VP elicits its hydro-osmotic (water permeability increase) effect by a stimulation of cAMP synthesis which is catalyzed by adenylate cyclase within the VP-sensitive cells of mammalian nephron (1). While all experimental criteria for mediatory role of cAMP in VP action in mammalian kidney are now fulfilled (1, 4), it can only be surmised that subsequent steps involve cAMP-dependent protein phosphorylations, micro-

---

[1] 'Primary prostaglandins' are called here relatively stable biologically active prostaglandins described originally, i.e. PGE, PGA, PGF – as distinguished from other newly discovered products of AA metabolism such as PGEP, thromboxanes and prostacyclin (55).

tubules and microfilaments (1, 5); the exact sequence of these steps distal to the cAMP synthesis has not yet been fully elucidated.

Classical experimental approaches used thus far to study the PG effect on VP action involve use of exogenously applied PG and use of drugs which inhibit PG synthesis or PG action (6). All these approaches have certain limitations. One of the major limitations is the existence of an enormous number of structurally diverse PG, which differ substantially in their functional and biochemical effects, as well as in their metabolic stability (52). The PG synthesis inhibitors used to date, such as nonsteroidal anti-inflammatory drugs (NSAID) inhibit the very initial step in the whole cascade of biosynthetic pathways initiated by the PG synthetase (PG cyclo-oxygenase), an enzyme utilizing arachidonic acid (AA) as a substrate (7, 52). Antagonists of PG action, e.g. polyphloretin phosphates (PPP), exhibit a similar lack of specificity toward different types of PG (8). In addition to this, both PG antagonists and PG synthesis inhibitors may have additional pharmacological effects, unrelated to PG, thus complicating the interpretation of experiments done with the use of these compounds (9). Accordingly, experiments using these drugs should be examined in view of these limitations and possible pitfalls.

Functional Interactions between VP and PG

The far most direct and clear evidence that PG modulate the VP antidiuretic effect on water permeability in mammalian kidney was provided by studies on isolated microperfused collecting tubules by *Grantham and Orloff* (10). These investigators found that PGE_1 blunted the increase in hydraulic water permeability elicited by VP, but did not inhibit the increase elicited by exogenous cAMP or by the cAMP phosphodiesterase (cAMP-PDIE) inhibitor theophylline. In addition, they observed that PGE_1 alone was capable of increasing water permeability and potentiating the effect of theophylline (10). These fundamental observations indicated that at least some types of PG can interfere with VP action at the step of VP-dependent cAMP generation. The data also suggested that VP-PG interactions are probably not limited only to the simple inhibition of VP-sensitive adenylate cyclase.

Experiments in which PG were infused systemically or into one renal artery yielded results from which interaction of PG with VP in distal nephron is difficult to interpret. PG (mostly PGE) (11—14) invariably increased urine flow and free water clearance but also increased solute excretion and renal blood flow, caused inhibition of proximal tubular reabsorption and washout of cortico-medullary gradient of solutes (11—15). Direct inhibitory effect of PG on VP action on distal nephron cannot be clearly dissected from these latter effects.

PG synthesis inhibitors administered *in vivo,* such as indomethacin (INDO) or meclofenamate (MCLF), were found to potentiate the antidiuretic effect of exogenous VP both in experimental animals and in human subjects (16–20, 51). In these studies INDO and MCLF pretreatment did not cause detectable changes in blood pressure, glomerular filtration rate (GFR), total renal blood flow, or osmolal clearance, thus suggesting that the VP-potentiating effect may be associated with or caused by a block of PG synthesis within the renal medulla. Such results favor the view that endogenous PG antagonize the antidiuretic VP effect by interfering with VP action on distal nephron. However, the effect of the tested NSAID on intrarenal redistribution of blood flow and on the gradient of solutes in renal medulla were not excluded (15) as a principal or contributory cause of potentiation of VP-induced antidiuresis. Moreover, for reasons mentioned above, such experiments do not show whether putative action to inhibit VP effect is due to the effect of primary PG (e.g. PGE_2), their precursors such as PG endoperoxides (PGEP), or some other multiple products of AA metabolism (52). For example, infusion of AA actually decreased urine flow (21) – an effect opposite to that of primary prostaglandins (11–15).

Studies on Cellular and Subcellular Systems

The question of whether PG may influence VP-dependent cAMP formation in renal medulla was approached in numerous studies (1, 22); the effects of PG on both tissue levels as well as on the enzymes of cAMP metabolism and action were examined. It should be stressed that in all these studies, without exception, experiments were performed on tissue slices or cell-free preparations from the whole medulla, medullopapillary, or papillary regions of the kidney. Renal medulla contains a number of other cell types (loops of Henle, interstitial cells, cells of vasa recta) in addition to the VP-sensitive cells. Such cell heterogeneity of the system makes clear-cut interpretations difficult.

Enzymes of cAMP Metabolism
Adenylate Cyclase
An obvious simple question – whether PG inhibit renal medullary adenylate cyclase (RMAC) – has not yet been satisfactorily resolved (1, 22). In some earlier studies, small inhibition of RMAC by primary PG was reported (23–25); in other studies no effect was found (26, 37). Still in other studies, primary PG actually stimulated RMAC, mostly in concentrations $>10^{-6}M$ (25, 26, 28, 29). Stimulation of RMAC with primary PG is much less extensive than stimulation with VP (1, 22) and results are more variable. Stimulation of RMAC from rabbit medullopapillary region appears to be dose-dependent and is inhibited by PPP (30). Stimulatory effect of VP and PG on RMAC preparations of human and

bovine kidney appear, to be additive (27, 28), which suggests that PG-sensitive and VP-sensitive RMAC are two different enzymes either in two different compartments of the same cell or in two distinct cell types both located within renal medulla.

It was recently shown for nonrenal systems that precursors (7, 52) or primary PG, namely PGEP, have quite an opposite effect on adenylate cyclase than primary PG (7). In adipocytes, PGH_2 inhibited the adenylate cyclase stimulated by several lipolytic hormones (7, 31). In thrombocytes, PGG_2 (32) or PGH_2 (33) actually inhibited the increase in cAMP caused by PGE_2 or by prostaglandin PGI_2 (34).

To test the possible involvement of PGEP in our recent experiments, we examined the effects of metabolically stable chemical analogues of PGH_2, namely, 15(s)-hydroxy, 9α–11α (epoxymethano)prosta-5,13-dienoic acid (11-Me-PGEX), or by 15(s)-hydroxy-hydroxy 11α,9α (epoxymethano)prosta-5,13-dienoic acid (9-Me-PGEX), on some components of cAMP system of rabbit renal papilla (37). We tested the effects of PGE_2 AA, 9-Me-PGEX and 11-Me-PGEX on adenylate cyclase *in vitro*, but none of these compounds had a significant effect on either basal or VP-stimulated activity of the enzyme (37).

Cyclic AMP Phosphodiesterase

cAMP-PDIE is another key enzyme controlling cAMP metabolism. The effect of PG on cAMP-PDIE has been much less extensively studied than the effect on adenylate cyclase. In epithelial tissues other than kidney medulla, such as gastric mucosa, primary PG exhibited little or no direct effect on cAMP-PDIE (35, 36), PGE_2 caused slight inhibition of membrane-bound canine fundic mucosa cAMP-PDIE (36). We examined recently the effect of various PG on cAMP-PDIE either contained in cytosol or membrane fraction of rabbit renal papilla. In a concentration range of 10^{-7} to 10^{-4} M PGE_2, AA, 9-Me-PGEX or 11-Me-PGEX had no effect on the activity of cAMP-PDIE (37).

Tissue Levels of cAMP in Renal Medulla

Disparity of results with respect to the effects of PG on renal medullary cAMP levels is similar to the results with RMAC in a cell-free system. PG alone (PGE_2) can cause a dose-dependent increase in renal medullary tissue cAMP both in rat or rabbit (30); this effect is blocked by PG inhibitor PPP (30). In inner medulla of rat kidney, PGE_1 was claimed to decrease tissue cAMP elevation induced by submaximal doses of VP (25). In inner medulla of dog (29) or rabbit (37), such an effect was not observed. In rat outer medulla the effects of PG and VP were reported to be additive (25).

In experiments using tissue slices of renal medulla or papilla, inhibitors of PG synthesis were sometimes used in an attempt to examine endogenous PG synthesis on cAMP levels and on the tissue response to VP. Pretreatment of rats with INDO (12) did not cause change in basal cAMP levels in renal medulla but increased the extent of the tissue cAMP elevation in response to VP (12). In another study, pretreatment of rat inner medullary slices with INDO and MCLF *in vitro* caused a lowering in basal inner cAMP levels. However, VP increased cAMP to the same level as in untreated control tissue (38). Recently we examined the effect not only of PGE_2 but also of PGEP analogues on tissue levels of cAMP (control or stimulated by VP) in renal papillary slices from control rabbits or rabbits pretreated with 9 mg/kg MCLF. VP elevated tissue cAMP in slices from control animals. PGE_2, 9-Me-PGEX or 11-Me-PGEX had no effect on cAMP levels alone. PGE_2 did not block increase in tissue cAMP by VP; increases in tissue cAMP by VP in presence of PGEP analogues were inconsistent but were not abolished (30).

Different responses were obtained on slices from MCLF-treated rabbits. Basal levels of renal papillary cAMP were significantly lower and response to VP was blunted. When added alone, 9-Me-PGEX and to a lesser degree 11-Me-PGEX or PGE_2 decreased the basal cAMP level, but relative tissue cAMP increases in response to VP were restored in the presence of both PGE_2 and PGEP analogues (37). These results would not support a thesis that PG (including PGEP) antagonize responsiveness to VP at the level of VP-dependent cAMP formation. On the contrary, it seems that the presence of endogenous PG (controls) or added exogenous PG in MCLF-treated animals rather favored the response to VP in terms of cAMP accumulation in renal papillary tissue.

Effects on cAMP-Dependent Protein Kinase (PK)

There are only a few observations on the possible effect of PG on PK. VP activates PK *in situ* in a dose-dependent way (39) and PK activation is likely one of the steps subsequent to cAMP formation (1, 4, 39). In preliminary experiments we found in bovine renal medullary slices that $10^{-7} M$ PGE_2 (39) increased the *in situ* activity of PK and this effect appeared additive to that of VP (39).

This finding indicates lack of inhibitory effect of exogenous PG also at PK activation step. Recently we examined the effect of pretreatment with MCLF on PK activity in renal papilla. State of PK activation, assessed by ($-$cAMP/+cAMP) PK activity ratio (4, 39), was significantly lower in MCLF-treated papillary tissue, probably reflecting lower tissue levels of cAMP. The extent of stimulation of the PK activity contained in 40,000 g supernate of homogenate by exogenous cAMP was significantly higher in MCLF-treated tissues — an observation deserving further analysis.

VP and PG Synthesis

Quite recently, a new type of relationship appears to be emerging between VP, cAMP and PG. It was found both in rabbit renal medullary interstitial cells in tissue culture (40) and in toad bladder (41) that VP can cause stimulation of PG synthesis, probably by activation of AA release from cellular phospholipids (40, 41). cAMP has no such effect (41); in some nonrenal systems cAMP actually appears to decrease the availability of AA for PG synthesis (42). Observations in rats (43) and rabbits (44) *in vivo* also suggest that VP stimulates the PG synthesis in kidney. The exact molecular mechanism by which VP stimulates PG synthesis remains to be elucidated. This newly discovered VP-PG relationship may represent an interesting regulatory pathway in the modulation of the renal antidiuretic response to VP at cellular level. Moreover, VP may indirectly VIA stimulation of PG synthesis in the kidney elicit renal effects, such as natriuresis, which were ascribed previously to the direct action of VP (51).

Conclusions and Perspectives

It is difficult, if not impossible, at the present time to offer a clear generalized scheme depicting exactly the mechanism by which intrarenal PG modulate the hydro-osmotic (antidiuretic) effect of VP.

All results obtained *in vivo* do support the notion that the increase in synthesis of endogenous renal medullary PG tends to attenuate, and the decrease in PG synthesis tends to accentuate, the antidiuretic effect of VP. Since VP itself might promote PG synthesis, current experimental evidence seems quite consistent with the original proposition by *Grantham and Orloff* (10) that PG may serve to dampen the alteration of water permeability of distal nephron resulting from small changes in peritubular concentrations of VP (10).

On the other hand, the cellular mechanism of such putative action of PG remains unclear. Most of the experimental evidence indicates that neither primary PG nor some analogues of PGEP decrease the VP-sensitive RMAC activity by direct inhibition of the enzyme. Likewise, cAMP-PDIE does not appear to be the site of PG action. It seems that there are two areas which deserve special attention in future studies aimed to elucidate the molecular basis of the modulatory effect of PG on VP in mammalian kidney. First, it should be explored whether some of the numerous products of intrarenal metabolism of AA, such as PGEP, thromboxanes or prostacyclines, are the major regulatory agents of the VP system, rather than primary PG. Second, it should be seriously explored whether PG and NSAID may act on steps distal to VP-dependent cAMP formation, such as cAMP-dependent protein phosphorylations.

In terms of the possible clinical and pathological significance of VP—PG interplay, first applications appear to emerge in relation to the VP-resistant renal concentrating defects. Reports from at least two centers suggest that in nephrogenic diabetes insipidus caused by potassium depletion, the urinary concentrating defect could be at least in part due to overproduction of PG. VP-resistant polyuria observed in potassium depletion was accompanied by high urinary PG excretion (45, 46) and treatment with INDO (45, 46) or MCLF (46) improved renal concentrating ability (45). Likewise, according to several case reports, NSAID were successfully used in the treatment of some other urinary concentrating defects, such as congenital nephrogenic diabetes insipidus (47) or VP-resistant polyuria caused by administration of lithium (48). In an experimental study in rats, urinary concentrating defect in enterococcal pyelonephritis was improved by administration of INDO or MCLF (49). Since inflammation could cause release of PG (50), it is not unreasonable to expect that urinary concentrating defects in renal diseases which involve renal medullary tissue may be caused by local PG overproduction and could be treated by agents which would inhibit the endogenous production of PG.

Summary

Current experimental evidence indicates that endogenous renal medullary prostaglandins modulate the antidiuretic response to vasopressin in the mammalian kidney. The predominant effect of prostaglandins is to attenuate the antidiuretic response to vasopressin; inhibition of prostaglandin synthesis potentiates the renal effect of vasopressin. Prostaglandins likely antagonize the renal effects of vasopressin at the cellular level of hormone-dependent cyclic adenosine 3,5-monophosphate metabolism, but the exact molecular mechanism is not known. Likewise, it is not known whether such modulatory effect is due to primary prostaglandins, prostaglandin precursors or to other metabolites of arachidonic acid. Vasopressin itself could stimulate intrarenal prostaglandin synthesis; this effect may represent a negative-feedback regulatory pathway for the antidiuretic response to the hormone. Recent experimental evidence suggests that modulatory effect of prostaglandin may be a factor in pathogenesis of some types of urinary concentrating defects.

Acknowledgments

Supported by NIH grant AM-16105 and by Mayo Foundation. Dr. *Thomas P. Dousa* is an Established Investigator of the American Heart Association. Mrs. *Ardith Walker* provided excellent secretarial assistance.

References

1 *Dousa, T. and Valtin, H.:* Cellular actions of vasopressin in the mammalian kidney. Kidney int. *10:* 46–63 (1976).
2 *Crowshaw, K. and Szlyk, J.:* Distribution of prostaglandins in rabbit kidney. Biochem. J. *116:* 421–424 (1970).
3 *Kuehl, F., jr.:* Prostaglandins, cyclic nucleotides and cell function. Prostaglandins *5:* 325–340 (1974).

4   *Hall, D.; Barnes, L., and Dousa, T.:* Cyclic AMP in action of antidiuretic hormone.
    Effects of exogenous cyclic AMP and its new analogue. Am. J. Physiol. *232:*
    F368–F376 (1977).

5   *Dousa, T.; Barnes, L., and Kim, J.:* The role of cyclic AMP-dependent protein phos-
    phorylations and microtubules in the cellular action of vasopressin in mammalian
    kidney; in *Moses and Share* Neurohypophysis, pp. 220–235 (Karger, Basel, 1977).

6   *Gryglewski, R.:* Structure-activity relationships of some prostaglandin synthetase in-
    hibitors; in *Robinson and Vane* Prostaglandin synthetase inhibitors, pp. 33–52 (Raven
    Press, New York 1974).

7   *Gorman, R.:* Prostaglandin endoperoxides. Possible new regulators of cyclic nucleo-
    tide metabolism. J. cyclic Nucl. Res. *1:* 1–9 (1975).

8   *Eakins, K.:* Prostaglandin antagonism by polymeric phosphates of phloretin and related
    compounds. Ann. N.Y. Acad. Sci. *180:* 387–395 (1971).

9   *Ferreira, S. and Vane, J.:* New aspects of the mode of action of nonsteroid anti-inflam-
    matory drugs. A. Rev. Pharmac. *14:* 57–73 (1974).

10 *Grantham, J. and Orloff, J.:* Effect of prostaglandin E_1 on the permeability response of
    the isolated collecting tubule to vasopressin, adenosine $3',5'$-monophosphate and
    theophylline. J. clin. Invest. *47:* 1154–1161 (1968).

11 *Shimizu, K.; Kurosawa, T.; Maeda, T., and Yoshitoshi, Y.:* Free water excretion and
    washout of renal medullary urea by prostaglandin E_1. Jap. Heart J. *10:* 437–455
    (1969).

12 *Martinez-Maldonado, M.; Tsaparas, M.; Eknoyan, G., and Suki, W.:* Renal actions of
    prostaglandins. Comparison with acetylcholine and volume expansion. Am. J. Physiol.
    222: 1147–1152 (1972).

13 *Gross, J. and Bartter, F.:* Effects of prostaglandins E_1, A and $F_{2\alpha}$ on renal handling of
    salt and water. Am. J. Physiol. *225:* 218–224 (1973).

14 *Berl, T. and Schrier, R.:* Mechanism of effect of prostaglandin E_1 on renal water
    excretion. J. clin. Invest. *52:* 463–471 (1973).

15 *Anderson, R.; Berl, T.; McDonald, K., and Schrier, R.:* Prostaglandins. Effects on blood
    pressure, renal blood flow, sodium and water excretion. Kidney int. *10:* 205–215
    (1976).

16 *Anderson, R.; Berl, T.; McDonald, K., and Schrier, R.:* Evidence for an *in vivo*
    antagonism between vasopressin and prostaglandin in the mammalian kidney. J. clin.
    Invest. *56:* 420–426 (1975).

17 *Silverstein, M.; Feldman, R.; Henderson, L., and Engelmann, K.:* Effects of in-
    domethacin on human renal clearance of sodium and H_2O. Clin. Res. *22:* 721A
    (1974).

18 *Lum, G.; Aisenbrey, G.; Dunn, M.; Berl, T.; Schrier, R., and McDonald, K.:* In
    vivo effect of indomethacin to potentiate the renal medullary cyclic AMP response
    to vasopressin. J. clin. Invest. *59:* 8–13 (1977).

19 *Berl, T.; Ras, A.; Wald, A.; Harowitz, L., and Czaczkes, W.:* Prostaglandin syn-
    thesis inhibition and the action of vasopressin. Studies in man and rat. Am. J.
    Physiol. *232:* F529–F537 (1977).

20 *Fröhlich, J.; Kokko, J.; Edwards, B., and Fulcher, S.:* Effect of inhibition of cycloxy-
    genase on water metabolism in man. Clin. Res. *25:* 432A (1977).

21 *Weber, P.; Holzgreve, H.; Stephan, R., and Herbst, R.:* Plasma renin activity and renal
    sodium and water excretion following infusion of arachidonic acid in rats. Eur. J.
    Pharmacol. *34:* 299–304 (1975).

22 *Dousa, T.:* Drugs and other agents affecting the renal adenylate cyclase system.
    Methods in Pharmacology *4A:* 293–331 (Plenum Press, New York 1976).

23 *Marumo, F. and Edelman, I.:* Effects of Ca^{++} and prostaglandin E_1 on vasopressin activation of renal adenyl cyclase. J. clin. Invest. *50:* 1613–1620 (1971).

24 *Kalisker, A. and Dyer, C.:* Inhibition of the vasopressin-activated adenyl cyclase from renal medulla by prostaglandins. Eur. J. Pharmacol. *20:* 143–146 (1972).

25 *Beck, N.; Kaneko, T.; Zor, U.; Field, J., and Davis, B.:* Effects of vasopressin and prostaglandin E_1 on the adenyl cyclase-cyclic 3′,5′-adenosine monophosphate system of the renal medulla of the rat. J. clin. Invest. *50:* 2461–2465 (1971).

26 *Schultz, G.; Jakobs, K.; Bohme, E. und Schultz, K.:* Einfluss verschiedener Hormone auf die Bildung von Adenosine-3′,5′-monophosphat und Guanosin-3′,5′-monophosphat durch partikulare Präparationen aus der Rattenniere. Eur. J. Biochem. *24:* 520–529 (1972).

27 *Dousa, T.P.:* Effect of prostaglandins on adenylate cyclase from human renal medulla; in *Kahn and Lands* Prostaglandins and cyclic AMP, pp. 155–156 (Academic Press, New York 1973).

28 *Birnbaumer, L. and Yang, P.:* Studies on receptor-mediated activation of adenylyl cyclases. Part I. J. biol. Chem. *249:* 7848–7856 (1974).

29 *Beck, N.; Reed, S.; Murdaugh, H., and Davis, B.:* Effects of catecholamines and their interaction with other hormones on cyclic 3′,5′-adenosine monophosphate of the kidney. J. clin. Invest. *51:* 939–944 (1972).

30 *Dousa, T.; Hui, Y.; Kim, J., and Northrup, T.:* Unpublished observations.

31 *Gorman, R.; Hamberg, M., and Samuelsson, B.:* Inhibition of basal and hormone stimulated adenylate cyclase in adipocyte ghosts by the prostaglandin endoperoxide prostaglandin H_2. J. biol. Chem. *250:* 6460–6463 (1975).

32 *Malmsten, C.; Granström, E., and Samuelsson, B.:* Cyclic AMP inhibits synthesis of prostaglandin endoperoxide (PGG_2) in human platelets. Biochem. biophys. Res. Commun. *68:* 569–576 (1976).

33 *Miller, O.; Johnson, R., and Gorman, R.:* Inhibition of PGE_1-stimulated cAMP accumulation in human platelets by thromboxane A_2. Prostaglandins *13:* 599–609 (1977).

34 *Gorman, R.; Bunting, S., and Miller, O.:* Modulation of human platelet adenylate cyclase by prostacyclin (PGX). Prostaglandins *13:* 377–388 (1977).

35 *Wollin, A.; Code, C., and Dousa, T.:* Interaction of prostaglandins and histamine with enzymes of cyclic AMP metabolism from guinea pig gastric mucosa. J. clin. Invest. *57:* 1548–1553 (1976).

36 *Dozois, R.; Kim, J., and Dousa, T.:* Interaction of prostaglandins with canine gastric mucosa adenylate cyclase-cyclic AMP system. Am. J. Physiol. (submitted, 1977).

37 *Northrup, T.; Kim, J.; Hui, Y., and Dousa, T.:* Interactions between prostaglandins (PG) and vasopressin (VP) in rabbit renal papilla. Physiologist *20:* 68 (1977).

38 *DeRubertis, F.; Zenser, T.; Craven, P., and Davis, B.:* Modulation of the cyclic AMP content of rat renal inner medulla by oxygen. J. clin. Invest. *58:* 1370–1378 (1976).

39 *Dousa, T. and Barnes, L.:* Regulation of protein kinase by vasopressin in renal medulla *in situ.* Am. J. Physiol. *231:* F50–F57 (1977).

40 *Zusman, R. and Keiser, H.:* Prostaglandin E_2 biosynthesis by rabbit renomedullary interstitial cells in tissue culture. J. biol. Chem. *252:* 2069–2071 (1977).

41 *Zusman, R.; Keiser, H., and Handler, J.:* Vasopressin-stimulated prostaglandin E biosynthesis in toad urinary bladder. Effect on water flow. J. clin. Invest. *60:* 1348–1353 (1977).

42 *Minkes, M.; Standford, N.; Chi, M.; Roth, G.; Raz, A.; Needleman, P., and Majerus, P.:* Cyclic adenosine 3′,5′-monophosphate inhibits the availability of arachidonate to prostaglandin synthetase in human platelet suspensions. J. clin. Invest. *59:* 449–454 (1977).

43 *Walker, L.; Whorton, R.; France, R.; Smigel, M., and Frolich, J.:* Antidiuretic hormone increases renal prostaglandin E_2 production in rats (Brattleboro) with hereditary hypothalamic diabetes insipidus. Fed. Proc. Fed. Am. Socs exp. Biol. *36:* 402 (1977).

44 *Lifschitz, M. and Stein, J.:* Antidiuretic hormone stimulates renal prostaglandin E (PGE) synthesis in rabbit. Clin. Res. *25:* 440A (1977).

45 *Galvez, O.; Roberts, B.; Bay, W., and Ferris, T.:* Studies of the mechanism of polyuria with hypokalemia. Clin. Res. *24:* 554A (1976).

46 *Youngberg, S.; Marchand, G.; Haas, J.; Romero, J., and Knox, F.:* Role of prostaglandins (PG) in nephrogenic diabetes insipidus (DI) like state following prolonged administration of mineralocorticoids. Clin. Res. *25:* 509A (1977).

47 *Fichman, M.; Speckart, P.; Zia, P., and Lee, A.:* Antidiuretic response to prostaglandin (PG) inhibition in primary (DI) and nephrogenic diabetes insipidus (NDI). Clin. Res. *25:* 505A (1977).

48 *Rutecki, G.W.; Nally, J.V.; Bay, W.H., and Ferris, T.F.:* The acute effects of lithium (Li) on renal function. Kidney int. *12:* 571 (1977).

49 *Levison, S. and Levison, M.:* Effect of indomethacin and sodium meclofenamate on the renal concentrating defect in experimental enterococcal pyelonephritis in rats. J. Lab. clin. Med. *88:* 958–964 (1976).

50 *Vane, J.:* Prostaglandins as mediators of inflammation. Adv. Prostagl. Thomboxane Res. *2:* 791 (1976).

51 *Fejes-Tóth, G.; Magyar, A., and Walter, J.:* Renal response to vasopressin after inhibition of prostaglandin synthesis. Am. J. Physiol. *232:* F416–F423 (1977).

52 *Samuelsson, B.:* Introduction. New trends in prostaglandin research. Adv. Prostagl. Thromboxane Res. *1:* 1–6 (1976).

53 *Zenser, T.; Levitt, M., and Davis, B.:* Effect of oxygen and solute on PGE and PGF production by rat kidney slices. Prostaglandins *13:* 143–151 (1977).

54 *Frölich, J.; Williams, W.; Sweetman, B.; Suigel, M.; Cern, K.; Hollifield, J.; Fisher, S.; Nies, A.; Frish-Holmberg, M., and Oates, J.:* Analysis of renal prostaglandin synthesis by competitive protein binding assay and gas chromatography – mass spectrometry. Adv. Prostagl. Thromboxane Res. *1:* 65–80 (1976).

55 *Dunn, M.J. and Hood, V.L.:* Prostaglandins and kidney. Am. J. Physiol. *233:* F169–F174 (1977).

Thomas P. Dousa, MD, PhD, Professor of Medicine and Physiology, Mayo Medical School, Mayo Clinic and Mayo Foundation, *Rochester, MN 55901* (USA)

The Renal Kallikrein-Kinin System

Contr. Nephrol., vol. 12, pp. 116–125 (Karger, Basel 1978)

The Kallikrein-Kinin System in the Kidney

John J. Pisano, Jenny Corthorn, Kerin Yates and Jack V. Pierce

Section on Physiological Chemistry, Laboratory of Chemistry, National Heart, Lung, and Blood Institute, The National Institutes of Health, Bethesda, Md.

Introduction

In 1909 it was observed that urine contains hypotensive as well as hypertensive substances (1). Intravenous injection of an alcohol-insoluble fraction of human urine equivalent to 80 ml urine caused a pronounced fall in the systolic pressure of the dog. The pressure remained depressed for 8 min and returned to normal in 13 min. The hypotensive action of urine also was noted in another laboratory a few years later (2). Working on the assumption that the active principle was a circulating hormone, a search was made for the producing gland (3). A large amount of the hypotensive substance was found in the pancreas and this organ was thought to be a source of the principle in urine (4). It was named kallikrein, a word derived from the Greek for pancreas. The first study relating urinary kallikrein to essential hypertension was performed by *Elliot and Nuzum* (5). They found urinary kallikrein was subnormal in hypertensive subjects, a finding confirmed a few years later (6). Interest in the kallikrein-kinin system in human hypertension lay dormant for 37 years until *Margolius et al.* (7) confirmed the finding of *Elliot and Nuzum* (5). Urinary kallikrein is subnormal in patients with essential hypertension (8) and supranormal in patients with primary aldosteronism (9). Urinary kallikrein in man and rat increases in response to a low-sodium diet or administration of sodium-retaining steroids. The aldosterone antagonist spironolactone abolishes the increase seen with low-sodium diets and primary aldosteronism (10–12). Kallikrein is also correlated with renal blood flow (13).

Urinary kallikrein comes from the kidney and not the pancreas. The enzyme in urine appears identical to the kallikrein isolated from kidney or synthesized in kidney slices (14). Urinary kallikrein may be correctly renamed renal kallikrein.

Early studies on the role of the kallikrein-kinin system in the kidney noted that kinins are potent vasodilators and cause natriuresis and diuresis when in-

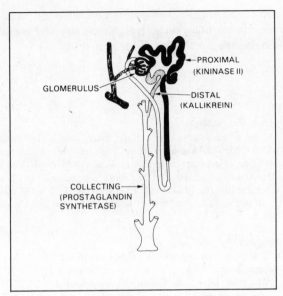

Fig. 1. Location of kallikrein and kininase in the nephron.

jected into the renal artery (15, 16). However, these interesting observations may have little relevance to normal occurrences in the kidney. Injected brady-kinin does not appear in urine (17, 18) because it is rapidly destroyed in the proximal (but not the distal) tubule (19) by kininase II (20) also known as the angiotensin converting enzyme.

Stop-flow studies in the dog indicate that renal kallikrein is secreted into the urine at the level of the distal tubule (21). In agreement with this observation is the immunohistochemical localization of the enzyme in the tubule (fig. 1), especially in the region from the macula densa to the collecting duct (22). Unlike the situation with renin, renal kallikrein has not been detected in blood, a finding consistent with the tubular localization of kallikrein and the localization of renin in the wall of the glomerular afferent arteriole. Nonetheless, low levels of kallikrein have been detected in perfusates of isolated rat kidney (23) and renin occurs in urine.

It is reasonable to suppose that a major site of kinin production is in the distal tubule and collecting duct and that kinin then stimulates the rich supply of prostaglandin synthetase observed in the collecting duct (24).

To better understand the role of the kallikrein-kinin system in the kidney, it is necessary to identify and localize all the components of the system and determine their interaction and turnover. In support of the concept of kinin generation in the nephron, we now report the presence of kininogen in urine. We also report ·on the presence of prokallikrein, the relative amounts of the three kinins, bradykinin, lysyl-bradykinin and methionyl-lysyl-bradykinin and, finally, the inactivation of kinins in urine.

Methods

Active and Inactive Kallikrein (Prokallikrein)

Kallikrein was determined by a radiochemical esterolytic assay using [³H]-Nα tosyl-arginine methyl ester ([³H]TAME) (25). Total kallikrein (active kallikrein plus prokallikrein) was determined by incubating 10 μl of gel-filtered urine with 30 μl 0.2 M Tris-HCl buffer pH 8.0 and 10 μl trypsin (1 μg). After 20 min, 10 μl of lima bean trypsin inhibitor (20 μg) was added and the solution was incubated for 30 min with 10 μl [³H]TAME (9 × 10⁴ dpm). Active kallikrein was determined in the absence of trypsin and prokallikrein was obtained by difference.

DEAE-Cellulose Chromatography

Dialyzed and concentrated urine was treated with 0.2 M NaCl to precipitate the Tamm and Horsfall glycoprotein. A sample, 1,219 A$_{280}$ units, was applied to a 2.5 × 38 cm column of DEAE-cellulose (Whatman DE-52 microgranular) equilibrated with 0.01 M phosphate, pH 6.0. Kallikrein was eluted with 1.5 liter of a linear gradient of 0.1–0.35 M phosphate, pH 6.0. Fraction volume 7.0 ml, flow rate 50 ml/h.

Urinary Kinins

Urinary kinins were separated on a SP-Sephadex C-25 column as previously described (26, 27) and bioassayed with the guinea pig ileum (28). Kinins were also determined by a radioimmunoassay procedure developed in our laboratory (*P. Haluska et al.*, to be published).

Urokininogen

Urine collected in plastic containers was immediately boiled for 15 min. To 2.0 ml boiled urine was added 1.8 ml 0.02 M Tris, 0.15 M NaCl buffer, pH 8.0. The solution was adjusted to pH 7.0 with HCl and 0.4 mg TPCK trypsin (200 μl of a 2 mg/ml solution containing 0.1 M Tris, 0.025 M CaCl$_2$, pH 7.0) was added. Samples were incubated 1 h at 37 °C, boiled for 12 min and kinins determined by radioimmunoassay. Urokininogen disappearance was determined by incubating the urine immediately after voiding for 3 h at 37 °C. A drop of toluene was added to inhibit microbial growth. Incubations were terminated by placing the samples in a boiling water bath for 6 min.

Kallikrein Recovery

Purified human urinary kallikrein, 0.2 mTAME units, was added to 20 μl urine and the solution was incubated for 80 min at room temperature. Kallikrein was then assayed as above.

Kinin Disappearance

Immediately after voiding, urine was cooled on ice and NaN$_3$ was added to make a 0.025% solution. Samples were incubated at 37 °C for 0, 1, 2 and 4 h. The reaction was stopped by placing the samples in a boiling water bath for 6 min. Kinins were determined by radioimmunoassay.

Urine Collection for Kinins
24-hour urine specimens were obtained from 17 normal subjects, 10 men and 7 women, ages 18–48 years. Urine was collected in plastic bottles containing 20 ml 6 N HCl (final pH 2–3) and stored at 4 °C.

Results and Discussion

Prokallikrein in Urine
When 10 urine samples obtained from healthy adults were incubated with trypsin, the level of kallikrein increased 57 ± 5% (mean ± SEM) (table I). No sex difference was observed. The trypsin-activatable kallikrein probably is prokallikrein, as it could be separated from active kallikrein on a DEAE-cellulose column (fig. 2) and has an apparent molecular weight of 50,000 daltons as compared with 35,000 for active kallikrein (29).

Kinins in Urine
Urine collected in acid contains three kinins, i.e., bradykinin, lysyl-bradykinin and methionyl-lysyl-bradykinin (27). The major kinin in men's urine was methionyl-lysyl-bradykinin and in women's urine, lysyl-bradykinin (table II). Other studies in our laboratory had shown that methionyl-lysyl-bradykinin is produced when purified kininogens are incubated with porcine pepsin (30) and it is highly probably that this urinary kinin was largely generated by uropepsin after acidification (27). Lysyl-bradykinin, the kinin formed by renal kallikrein

Table I. Active and inactive kallikrein esterase activity in normal human urine

Subject	Sex	Free	mTU/ml inactive	Total	% inactive
1	M	29.6	10.9	40.5	27
2	F	4.6	3.0	7.6	40
3	F	19.3	15.9	35.2	45
4	M	10.1	12.6	22.7	56
5	M	17.2	22.0	39.2	56
6	F	7.7	11.8	19.5	60
7	M	3.9	6.5	10.4	63 ·
8	M	3.6	8.9	12.5	71
9	F	1.8	5.0	6.8	74
10	F	20.0	73.6	93.6	79
Mean ± SEM					57 ± 5

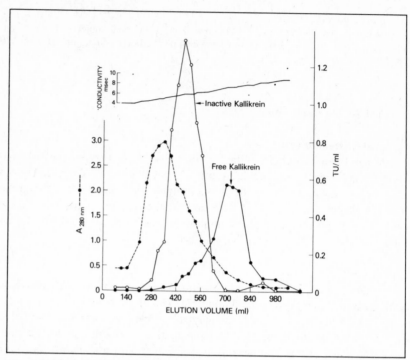

Fig. 2. Separation of prokallikrein (inactive kallikrein) and kallikrein by DEAE-cellulose chromatography (see 'Methods').

(31), is normally the major kinin in urine. About half is converted to bradykinin, presumably by the kinin-converting aminopeptidase (32) which has been isolated from human urine (*J.A. Guimarães and J.V. Pierce*, personal commun.). The significance of the three kinins and kinin conversion is totally unknown.

Urokininogen

The occurrence of kininogen in urine was indicated by the generation of methionyl-lysyl-bradykinin in urine after voiding. Kininogen antigen had been detected earlier in urine by the Ouchterlony immunodiffusion method using antibodies to plasma kininogens and a protein concentrate of human urine. A line of identity was obtained between the plasma and urine antigen. It was not known if the urine antigen represented degradation products of the plasma protein or if it contained kinin. When seven urine samples were incubated with trypsin, substantial kinin was generated in every case. The kinin content increased 61 ± 7% (mean ± SEM) (table III). Purified human urinary kallikrein added to urine also generated kinins.

Table II. Urinary kinins in $\mu g/24$ h

Subject	Bradykinin	Lysyl-bradykinin	Methionyl-lysyl-bradykinin	Total
Men				
1	2.7	11.6	22.0	36.3
2	1.4	13.0	13.9	28.3
3	7.5	9.6	9.9	27.0
4	5.2	10.6	7.6	23.4
5	2.9	4.2	10.3	17.4
6	0.9	8.8	6.3	16.0
7	4.0	4.0	7.2	15.2
8	1.2	5.7	7.9	14.8
9	4.1	4.0	5.4	13.5
10	4.3	4.5	4.3	13.1
Mean ± SEM	3.4 ± 0.6	7.6 ± 1.1	9.5 ± 1.6	20.5 ± 2.5
Women				
1	5.6	6.7	2.8	15.1
2	2.9	8.9	2.6	14.4
3	4.0	4.2	1.3	9.5
4	1.0	6.5	0.7	8.0
5	0.4	5.0	0.8	6.2
6	1.9	3.1	<0.1	5.0
7	1.7	1.3	0.3	2.5
Mean ± SEM	2.5 ± 0.7	5.1 ± 1.0	1.2 ± 0.4	8.6 ± 1.8

Table III. Kinin generation from urokininogen

Urine sample	Sex	Kinin		
		before trypsin ng/ml urine	after trypsin ng/ml urine	kinin increase %
1	F	3.6	5.0	39
2	F	12.4	18.2	47
3	M	3.2	4.8	50
4	F	3.8	6.0	58
5	F	13.8	23.2	68
6	M	7.8	13.2	69
7	M	8.0	15.6	95
Mean ± SEM				61 ± 7

Table IV. Urokininogen slowly decreases in incubated urine

Sample	Sex	pH	Active kallikrein mTU/ml urine	Free kinin ng/ml urine	Kinin in kininogen	
					µg/ml urine	decrease after incubation, %
1	M	6.4	6.2	25	2.0	0
2	F	5.3	7.2	12.0	9.4	2
3	F	5.5	8.9	23.0	17.0	29
4	M	5.5	3.2	3.9	35.0	35
5	M	5.7	3.6	13.0	7.0	40
6	F	6.3	7.9	33.0	17.0	74
7	F	6.7	0.85	0.64	1.0	100
8	M	6.1	0.40	0.36	1.6	100
Mean ± SEM						48 ± 14

Table V. Added urinary kallikrein is partially recovered from urine

Sample	pH	Recovery of esterase activity, %
1	5.7	44
2	6.0	44
3	5.8	60
4	6.3	68
5	5.6	69
6	6.3	70
7	6.3	72
8	6.2	82
Mean ± SEM		64 ± 5

Normally, kininogen is slowly hydrolyzed in urine. In a test of eight samples, kininogen decreased 48 ± 14% (mean ± SEM) after 3 h incubation without pH adjustment (table IV). This slow hydrolysis may be due in part to the partial inhibition of kallikrein in urine as only 64 ± 5% of human urinary kallikrein added to urine was recovered (table V).

Kinase
Kinins are not as rapidly destroyed in urine as in plasma. The half life of kinin in plasma is less than 30 sec, but in urine it is greater than 2 h when urine was incubated without pH adjustment. In six specimens 38 ± 6% (mean ± SEM)

Table VI. Kinin disappearance in incubated urine

Sample	Sex	pH	Initial kinin ng/ml	Disappeance in 2 h, %
1	M	5.9	3.4	18
2	F	6.0	12.3	27
3	M	5.2	21.3	38
4	M	5.3	47.8	39
5	F	6.7	13.9	51
6	F	7.2	3.2	53
Mean ± SEM				38 ± 6

disappeared in 2 h. In 4 h, 41–82% disappeared. The highest rates of disappearance occurred in the urine samples closer to neutrality. No sex difference was evident (table VI).

Conclusion

The increasing number of reports on alterations in kallikrein excretion will become more meaningful when other components of the kallikrein-kinin system are also determined, including prokallikrein, renal kininogen, individual kinins, kininase(s), the kinin-converting aminopeptidase, kallikrein inhibitor(s) and the presumed prokallikrein activating enzyme. Any one or all of these components could play a major role in regulating the level of kinin, the effector hormone. Studies in our laboratory have shown that there is usually no correlation between the level of kallikrein and kinins in urine (33). Perhaps the availability of renal kininogen and/or the activity of kininase(s) are major determinants of the level of kinins in the kidney.

Summary

To understand the role of the kallikrein-kinin system in the kidney all components of the system and their localization need to be considered. About half the kallikrein in urine occurs as the proenzyme which arises in the distal tubule. Kinins are formed in the distal tubule and collecting duct from urokininogen which is found throughout the tubule. Urine contains about twice as much lysyl-bradykinin as bradykinin. A third kinin, methionyl-lysyl-bradykinin, also can occur in urine. It is probably produced by uropepsin as the kinin is largely formed in acidified urine and its formation is inhibited by pepstatin. The significance of the three kinins is unknown. Kinins are normally slowly (few hours) destroyed in urine. The importance of kallikrein, urokininogen and kininases in regulating the level of kinins needs to be determined.

References

1 *Abelous, J.E. et Bardier, E.:* Les substances hypotensives de l'urine humaine normale. C.r. Séanc. Soc. Biol. *66:* 511–512 (1909).
2 *Frey, E.K.:* Zusammenhänge zwischen Herzarbeit und Nierentätigkeit. Arch. klin. Chir. *142:* 663 (1926).
3 *Frey, E.K.; Kraut, H. und Werle, E.:* Kallikrein (Padutin) (Enke, Stuttgart 1950).
4 *Frey, E.K.; Kraut, H.; Werle, E.; Vogel, R.; Zickgraf-Rüdel, G. und Trautschold, I.:* Das Kallikrein-Kinin-System und seine Inhibitoren (Enke, Stuttgart 1968).
5 *Elliot, A.H. and Nuzum, F.R.:* Urinary excretion of a depressor substance (kallikrein of Frey and Kraut) in arterial hypertension. Endocrinology *18:* 462–474 (1934).
6 *Werle, E. und Korsten, H.:* Der Kallikreingehalt des Harns, des Speichels und des Blutes bei Gesunden und Kranken. Z. ges. exp. Med. *103:* 153–162 (1938).
7 *Margolius, H.S.; Geller, R.; Pisano, J.J., and Sjoerdsma, A.:* Altered urinary kallikrein excretion in human hypertension. Lancet *ii:* 1063–1065 (1971).
8 *Margolius, H.S.; Horwitz, D.; Pisano, J.J., and Keiser, H.R.:* Urinary kallikrein excretion in hypertensive man. Relationships to sodium intake and sodium-retained steroids. Circulation Res. *35:* 820–825 (1974).
9 *Margolius, H.S.; Horwitz, D.; Geller, R.G.; Alexander, R.W.; Gill, J.R., jr.; Pisano, J.J., and Keiser, H.R.:* Urinary kallikrein excretion in normal man. Relationships to sodium intake and sodium-retaining steroids. Circulation Res. *35:* 812–819 (1974).
10 *Margolius, H.:* The kallikrein-kinin system and human hypertension; in *Pisano and Austen* Chemistry and biology of the kallikrein-kinin system in health and disease, pp. 399–409 (US Government Printing Office, Washington 1976).
11 *Geller, R.G.:* Urinary kallikrein excretion in normotensive and hypertensive rats; in *Pisano and Austen* Chemistry and biology of the kallikrein-kinin system in health and disease, pp. 379–387 (US Government Printing Office, Washington 1976).
12 *Croxatto, H.R.; Roblero, J.; Albertini, R.; Corhorn, J.; San Martin, M., and Porcelli, G.:* The kallikrein-kinin system in renal hypertension; in *Pisano and Austen* Chemistry and biology of the kallikrein-kinin system in health and disease, pp. 389–397 (US Government Printing Office, Washington 1976).
13 *Keiser, H.R.; Andrews, M.J., jr.; Guyton, R.A.; Margolius, H.S., and Pisano, J.J.:* Urinary kallikrein in dogs with constriction of one renal artery. Proc. Soc. exp. Biol. Med. *151:* 53–56 (1976).
14 *Nustad, K.; Vaaje, K., and Pierce, J.V.:* Synthesis of kallikreins by rat kidney slices. Br. J. Pharmacol. *53:* 229–234 (1975).
15 *Webster, M.E. and Gilmore, J.P.:* Influence of kallidin-10 on renal function. Am. J. Physiol. *206:* 714–718 (1964).
16 *Willis, L.R.; Ludens, J.H.; Hook, J.B., and Williamson, H.E.:* Mechanism of natriuretic action of bradykinin. Am. J. Physiol. *217:* 1–5 (1969).
17 *Yoshinaga, K.; Abe, K.; Miwa, I.; Furuyama, T., and Suzuki, C.:* Evidence for the renal origin of urinary kinin. Experientia *20:* 396–397 (1964).
18 *Nasjletti, A.; Colina-Chourio, J., and McGiff, J.C.:* Effect of kininase inhibition on canine renal blood flow and sodium excretion. Acta physiol. latinoam. *24:* 587–591 (1974).
19 *Carone, F.A.; Pullman, T.N.; Oparil, S., and Nakamura, S.:* Micropuncture evidence of rapid hydrolysis of bradykinin by rat proximal tubule. Am. J. Physiol. *230:* 1420–1424 (1976).
20 *Ward, P.W.; Gedney, C.D.; Dowben, R.M., and Erdos, E.G.:* Isolation of membrane-bound renal kallikrein and kininase. Biochem. J. *15:* 755–758 (1975).

21 *Scicli, A.G.; Carretero, O.A.; Hampton, A.; Cortes, P., and Oza, N.B.:* Site of kininogenase secretion in the dog nephron. Am. J. Physiol. *230:* 533–536 (1976).

22 *Ørstavik, T.B.; Nustad, K.; Brandtzaeg, P., and Pierce, J.V.:* Cellular origin of urinary kalikreins. J. Histochem. Cytochem. *24:* 1037–1039 (1976).

23 *Roblero, J.; Croxatto, H.R.; Corthorn, J.; Garcia, R.L., and DeVito, E.:* Kininogenase activity in urine and perfusion fluid of isolated rat kidney. Acta physiol. latinoam. *23:* 566–568 (1973).

24 *Smith, W.L. and Wilkin, G.P.:* Distribution of prostaglandin-forming cyclooxygenases in rat, rabbit and guinea pig kidney as determined by immunofluorescence. Abstract. Fed. Proc. Fed. Am. Socs exp. Biol. *36/3:* 309 (1977).

25 *Imanari, T.; Kaizu, T.; Yoshida, H.; Yates, K.; Pierce, J.V., and Pisano, J.J.:* Radiochemical assays for human urinary, salivary, and plasma kallikreins; in *Pisano and Austen* Chemistry and biology of the kallikrein-kinin system in health and disease, pp. 205–213 (US Government Printing Office, Washington 1976).

26 *Sampaio, M.U.; Reis, M.L.; Fink, E.; Camargo, A.C.M., and Greene, L.J.:* Chromatographic systems for desalting and separating kinins. Application to trypsin-treated human plasma. Abstract. Life Sci. *16:* 796 (1975).

27 *Hial, V.; Keiser, H.R., and Pisano, J.J.:* Origin and content of methionyl-lysyl-bradykinin, lysyl-bradykinin an; bradykinin in human urine. Biochem. Pharmac. *25:* 2499–2503 (1976).

28 *Trautschold, I.:* Assay methods in the kinin systems; in *Erdos* Handbook of experimental pharmacology, vol. 25, pp. 52–81 (Springer, Berlin 1970).

29 *Corthorn, J.; Imanari, T.; Yoshida, H.; Kaizu, T.; Pierce, J., and Pisano, J.:* Inactive kallikrein in human urine. Abstract. Fed. Proc. Fed. Am. Socs exp. Biol. *36/3:* 893 (1977).

30 *Guimaraes, J.A.; Pierce, J.V.; Hial, V., and Pisano, J.J.:* Methionyl-lysyl-bradykinin. The kinin released by pepsin from human kininogens; in *Sicuteri, Back and Haberland* Kinins: pharmacodynamics and biological roles, pp. 265–269 (Plenum Press, New York 1976).

31 *Pierce, J.V. and Webster, M.E.:* Human plasma kallidins. Isolation and chemical studies. Biochem. Biophys. Res. Commun. *5:* 353–357 (1961).

32 *Brandi, C.M.W.; Prado, E.S.; Prado, M.J.B.A., and Prado, J.L.:* Kinin-converting aminopeptidase from human urine partial purification and properties. Int. J. Biochem. *7:* 335–341 (1976).

33 *Vinci, J.; Zusman, R.; Bowden, R.; Horwitz, D., and Keiser, H.:* Relationship of urinary and plasma kinins to sodium retaining steroids and plasma renin activity. Abstract. Clin. Res. *25/3:* 450 A (1977).

Dr. *J.J. Pisano,* Section on Physiological Chemistry, Laboratory of Chemistry, National Heart, Lung, and Blood Institute, National Institutes of Health, *Bethesda, MD 20014* (USA)

Contr. Nephrol., vol. 12, pp. 126–131 (Karger, Basel 1978)

Effects of Aldosterone and Deoxycorticosterone on the Urinary Excretion of Kallikrein and of Prostaglandin E-Like Substance in the Rat

J. Colina-Chourio, J.C. McGiff and A. Nasjletti

Department of Pharmacology, University of Tennessee Center for the Health Sciences, Memphis, Tenn.

An interaction of kinins and renal prostaglandins was suggested by the demonstration that intrarenal arterial infusion of bradykinin evokes release of a prostaglandin E-like substance from canine kidney (8). More recently, we reported that kinins generated intrarenally selectively increase the output of prostaglandin E-like material from rabbit isolated kidney perfused with Krebs' solution (2). However, an influence of the renal kallikrein-kinin system on prostaglandin release is not established in intact animals. If the intrarenal activity of the kallikrein-kinin system influences kidney prostaglandin release, it may be expected that administration of mineralocorticoids, which increase urinary kallikrein (5), would promote renal prostaglandin release. To evaluate this possibility, we examined in the rat the effects of aldosterone and deoxycorticosterone (DOCA) on the urinary output of kallikrein and prostaglandin E-like substance (PGE). Urinary prostaglandins arise intrarenally and reflect renal prostaglandin release (3). Similarly, urinary kallikrein is synthesized by the kidney and presumably serves as indicator of the intrarenal activity of the kallikrein-kinin system (14).

Methods

Male Sprague-Dawley rats (300–350 g) were placed in metabolism cages and maintained on a commercial chow containing 152 mEq of sodium/kg and tap water. Steroids, in 0.2 ml of sesame oil, were given daily by subcutaneous injection for 14 days. The animals were divided into three groups: one group received deoxycorticosterone acetate (DOCA, 5 mg/day), the second, d-aldosterone acetate (0.25 mg/day) and the third, sesame oil (0.2 ml/day). 24-hour urine collections, obtained from individual rats before and 2 weeks after commencing steroid administration, were used to measure sodium (flame photometry), kallikrein and PGE-like substance. Kallikrein was measured by the method of *Marin-Grez et al.* (6) which is based on the determination of the amount of kinin formed when urine (5 μl) is incubated (20 min, 37 °C, pH 7.5) with an excess of kininogen in the presence of kininase inhibitors. The kinin formed was assayed in the canine hind-limb preparation using synthetic bradykinin (Sandoz) as reference standard (10). Kallikrein excretion, calculated by multiplying urinary kallikrein activity by urine flow, was expressed as micrograms of bradykinin equivalents per day.

The content of PGE-like substance in urine was determined as described previously (2) with minor modifications. Briefly, urine (10–15 ml) was combined with 5 vol of acetone, passed through a Millipore filter and evaporated under reduced pressure at 35 °C. This was followed by acidification (pH 3.5) with formic acid and extraction (four times) with an equal volume of ethyl acetate; acidic lipids in the combined ethyl acetate phases were subsequently separated from neutral lipids by successive extraction with $0.1\,M$ potassium phosphate buffer pH 8 and chloroform. The acidic lipids were purified further by chromatography on thin layers of silica gel (Silica Gel G plates, 0.5 mm thick, Brinkmann Instruments) using the solvent system chloroform:methanol:acetic acid (18:2:1, by volume). Marker plates, prepared by spotting 10 μg of authentic PGE_2 and $PGF_{2\alpha}$ were run concurrently which preparative plates. The zone of the preparative plate corresponding to the position of PGE_2 was scraped off and eluted with chloroform:methanol (4:1, v/v); the eluate was dried in N_2, reconstituted in $0.15\,M$ NaCl, and bioassayed for its content of PGE-like material using PGE_2 as reference standard. The concentration of PGE-like substance was determined by parallel pharmacological assay using the stomach and colon of rat and the rectum of chick, superfused in series with Krebs' solution, as assay organs (17). Excretion of PGE-like substance (24-hour urine volume × PGE concentration) was expressed as nanograms of PGE_2 equivalents per day. Plasma renin concentration, measured in blood obtained by heart puncture (0.5 ml) before and 2 weeks after commencing DOCA treatment, was expressed as nanograms of angiotensin II equivalents per milliliter of plasma per hour of incubation (12). Blood pressure was measured by tail sphygmography in unanesthetized animals after warming. All results, expressed as means ± SE were analyzed for significance with Student's t-test for paired observations.

Results

Administration of DOCA increased PGE and kallikrein excretion, measured after 14 days of treatment, from 36.9 ± 9.7 ng/day and 17.46 ± 4.27 μg/day to 103.6 ± 18.2 ng/day (p <0.02) and 65.57 ± 8.56 μg/day (p <0.005), respectively (table I). Concomitantly, the steroid reduced plasma renin from 20.5 ± 2.6 to 5.0 ± 1.4 ng ml^{-1} h^{-1} (p <0.005) and increased urine flow by about threefold (p <0.02); neither blood pressure nor sodium excretion were altered (table I). Aldosterone, given daily for 14 days, had similar effects; the steroid increased PGE excretion from 38.5 ± 7.1 to 95.4 ± 17.1 ng/day (p <0.02), kallikrein excretion from 18.50 ± 3.76 to 46.71 ± 8.64 μg/day (p <0.05), and urine flow about threefold (p <0.005); neither blood pressure nor sodium excretion were affected (table I). In contrast, in those rats that served as controls, the values for urine flow and excretion of sodium, PGE and kallikrein were not altered by administration of vehicle.

Table I. Effects of DOCA and of aldosterone on the urinary excretion of kallikrein and of PGE-like substance in the rat (mean ± SE)

Treatment	PGE excretion ng/day		Kallikrein excretion µg/day		Urine flow ml/day		Sodium excretion mEq/day		Blood pressure mm Hg	
	C	E	C	E	C	E	C	E	C	E
DOCA (5 mg/day; n = 6)	36.9 ± 9.7	103.6 ± 18.2	17.46 ± 4.27	65.57 ± 8.65	10.7 ± 1.8	38.7 ± 5.6	0.99 ± 0.88	1.21 ± 0.14	124 ± 4	125 ± 5
p		<0.02		<0.005		<0.02		NS		NS
Aldosterone (0.25 mg/day; n = 5)	38.5 ± 7.1	95.4 ± 17.1	18.50 ± 3.76	46.71 ± 8.64	11.3 ± 1.4	34.6 ± 3.3	1.47 ± 0.07	1.85 ± 0.24	114 ± 3	118 ± 7
p		<0.02		<0.05		<0.005		NS		NS
Oil (0.2 ml/day; n = 5)	36.8 ± 7.4	30.3 ± 3.7	20.82 ± 4.30	28.19 ± 3.15	10.5 ± 1.8	13.0 ± 2.3	1.52 ± 0.18	1.56 ± 0.34	110 ± 2	113 ± 2
p		NS		NS		NS		NS		NS

C and E indicate values obtained before and after 14 days of treatment; n = number of rats; p is derived from a t-test on paired values; NS indicates $p < 0.05$.

Discussion

The demonstration that aldosterone and DOCA increase kallikrein and PGE excretion suggests interactions of renal prostaglandins, mineralocorticoid hormones, and the renal kallikrein-kinin systems (9). These steroids, either directly or through the electrolyte and hormonal disturbances, i.e., K^+ depletion, kallikrein stimulation, produced by their continuous administration, may either stimulate the synthesis or decrease the inactivation of renal prostaglandins. Augmentation by mineralocorticoids of urinary PGE excretion is unrelated to alterations of the renin-angiotensin system produced by the steroids, i.e., reduced plasma renin activity and diminished generation of angiotensin II, since angiotensins I and II promote PGE release (3, 13). Demonstration that urinary output of PGE and kallikrein augmented in response to treatment with aldosterone and DOCA is consistent with the hypotheses that kinins generated intrarenally contribute to the effect of mineralocorticoids on urinary PGE. Our experiments in Krebs' perfused rabbit kidneys, showing reduction by a kallikrein inhibitor of the augmented PGE release evoked by kininogen, suggested stimulation of kidney PGE output by kinins generated intrarenally (2). It is possible that kallikreins synthesized in the kidney cortex (14), and released in urine at the level of the distal tubule (16), act on kininogen to produce kallidin which in turn reaches the sites of prostaglandin synthesis in the renal medulla (4) via the collecting duct, and evokes PGE release. This possibility is endorsed by the demonstration in urine of free kinins which originate intrarenally (11) and by the occurrence of kininogen in kidney (15).

Uninterrupted synthesis of prostaglandins by the kidney may influence renal blood flow, its intrarenal distribution, and sodium excretion (1). Similarly, kinins generated intrarenally appear to promote renal vasodilation, diuresis and natriuresis (9, 11). Mediation by prostaglandins of some of the renal actions of kinins may be possible (7). These observations and our present findings suggestive of interactions of mineralocorticoids, kinins, and prostaglandins, may help to explain the mechanism(s) of the sodium escape and of the polyuria associated with administration of mineralocorticoid hormones. Thus, enhanced intrarenal activity of the prostaglandin and kallikrein-kinin systems may facilitate the excretion of salt and water, thereby opposing the renal actions of sodium-retaining steroids.

Summary

Deoxycorticosterone (5 mg) and aldosterone (0.25 mg), given to rats for 14 days, increased the urinary excretion of kallikrein and of prostaglandin E-like substance and produced polyuria, but affected neither sodium excretion nor blood pressure. These results suggest that interactions of mineralocorticoid hormones, kinins and of prostaglandins may be important in the maintenance of salt-water homeostasis.

Acknowledgements

Supported by USPHS Grant HL-18579, American Heart Association Grant 76781, and by the Tennessee Heart Association. A.N. is the recipient of National Heart and Lung Institute Research Career Development Award 1 KO4 HL 00163.

References

1 *Anderson, R.J.; Berl, T.; McDonald, M.K., and Schrier, R.W.:* Prostaglandins. Effects on blood pressure, renal blood flow, sodium and water excretion. Kidney int. *10:* 205–215 (1976).

2 *Colina, J.; Miller, M.P.; McGiff, J.C., and Nasjletti, A.:* Dependency of prostaglandin release on intrarenal generation of kinins. Br. J. Pharmacol. *58:* 165–172 (1976).

3 *Frolich, J.C.; Wilson, T.W.; Sweetman, B.J.; Smigel, M.; Nies, A.S.; Carr, K.; Watson, J.T., and Oates, J.A.:* Urinary prostaglandins: identification and origin. J. clin. Invest. *55:* 763–770 (1975).

4 *Larsson, C. and Anggard, E.:* Regional differences in the formation and metabolism of prostaglandins in the rabbit kidney. Eur. J. Pharmacol. *21:* 30–36 (1973).

5 *Margolius, H.S.; Geller, R.G.; DeJong, W.; Pisano, J.J., and Sjoerdsma, A.:* Urinary kallikrein excretion in hypertension. Circulation Res. *30/31:* suppl. II, pp. 125–131 (1972).

6 *Marin-Grez, M.; Cottone, J., and Carretero, O.A.:* A method for measurement of urinary kallikrein. J. appl. Physiol. *32:* 428–431 (1972).

7 *McGiff, J.C.; Itskovitz, H.D., and Terragno, N.A.:* The actions of bradykinin and eledoisin in the canine isolated kidney: relationships to prostaglandins. Clin. Sci. molec. Med. *49:* 125–131 (1975).

8 *McGiff, J.C.; Terragno, N.A.; Malik, K.U., and Lonigro, A.J.:* Release of a prostaglandin E-like substance from canine kidney by bradykinin. Circulation Res. *31:* 36–43 (1972).

9 *Nasjletti, A. and Colina-Chourio, J.:* Interaction of mineralocorticoids, renal prostaglandins and the renal kallikrein-kinin system. Fed. Proc. Fed. Am. Socs exp. Biol. *35:* 189–193 (1976).

10 *Nasjletti, A.; Colina-Chourio, J., and McGiff, J.C.:* Assay of kinins by their effects on canine femoral blood flow. Proc. Soc. exp. Biol. Med. *150:* 493–497 (1975).

11 *Nasjletti, A.; Colina-Chourio, J., and McGiff, J.C.:* Disappearance of bradykinin in the renal circulation of dogs: effects of kininase inhibition. Circulation Res. *37:* 59–65 (1975).

12 *Nasjletti, A. and Masson, G.M.C.:* A method for the measurement of plasma renin in the rat. Proc. Soc. exp. Biol. Med. *136:* 344–348 (1971).

13 *Needleman, P.; Kauffman, A.H.; Douglas, J.R.; Johnson, E.M., and Marshall, G.R.:* Specific stimulation and inhibition of renal prostaglandin release by angiotensin analogs. Am. J. Physiol. *224:* 1415–1419 (1973).

14 *Nustad, K.; Vaaje, K., and Pierce, J.V.:* Synthesis of kallikreins by rat kidney slices. Br. J. Pharmacol. *53:* 229–234 (1975).

15 *Sardesai, V.M.:* Determination of bradykinin in blood and bradykininogen in tissues. Can. J. Physiol. Pharmacol. *46:* 77–79 (1968).

16 *Scicli, A.G.; Carretero, O.A.; Cortes, P., and Oza, N.B.:* Site of kininogenase secretion in the dog nephron. Am. J. Physiol. *230:* 533–536 (1976).

17 *Vane, J.R.:* Release and fate of vasoactive hormones in the circulation. Br. J. Pharmacol. *35:* 209–242 (1969).

Dr. *A. Nasjletti,* Department of Pharmacology, University of Tennessee, Center for the Health Sciences, 800 Madison Avenue CR-301, *Memphis, TN 38163* (USA)

Contr. Nephrol., vol. 12, pp. 132–144 (Karger, Basel 1978)

Stimulation of the Renal Kallikrein-Kinin System by Vasoactive Substances and Its Relationship to the Excretion of Salt and Water[1]

Ivor H. Mills, L.F.O. Obika and Pamela A. Newport

Department of Medicine, University of Cambridge, Addenbrooke's Hospital, Cambridge

Introduction

Since the original studies by *de Wardener et al.* (9) and *Mills et al.* (26) demonstrating the existence of a mechanism, other than changes in glomerular filtration or aldosterone secretion affecting sodium excretion, many groups of workers have sought for a natriuretic hormone. It has usually been assumed that a single substance would be identified. The concept has been mainly of a circulating substance which acted on the kidney to produce natriuresis though it was pointed out in 1970 (25) that the evidence supported the view that if there were a natriuretic hormone acting directly on renal tubular cells it would have to arise from the kidney itself.

With a variety of substances being referred to as possible contenders for the role of 'the' natriuretic hormone, it seemed more probable that there is a natriuretic hormone system and that a variety of stimulating or inhibiting actions would impinge on this chain of events at different points. Starting with the actions of angiotensin II it is now possible to define most of the links in the chain of hormonal events.

Natriuretic Action of Angiotensin II

Angiotensin II infusions have been reported as producing either sodium retention or natriuresis (2, 10, 32). This complex situation has been clarified by the infusion of angiotensin II directly into the renal artery in the dog (18). The infusion of the peptide at a low dosage (10–100 ng/min) leads to sodium

[1] This work was supported by the National Kidney Research Fund and the Medical Research Council. L.F.O.O. was a Junior Research Fellow of the University of Nigeria.

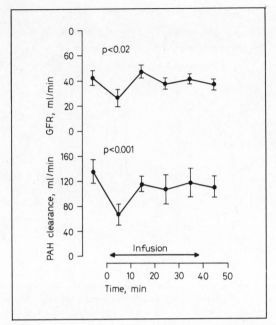

Fig. 1. The inulin clearance (GFR, ml/min) and PAH clearance (ml/min) prior to, during and following the intra-arterial infusion of angiotensin II at 10 μg/min in dogs. The first point is the mean and SE of two 20-min control periods and the last point is the mean of the 10-min period after stopping angiotensin. Each point is the mean and SE of six experiments. The p values represent the significance of the rise in clearances in the second infusion period. From *Klein et al.* (18).

retention, whereas at rates of 5–10 μg/min it is natriuretic. At intermediate dosages (1 μg/min) renal vasoconstriction is clearly seen, as shown by decrease in the clearances of inulin and p-aminohippurate (PAH). At the higher dosages the initial vasoconstriction decreases while the infusion continues, and clearances return to normal or even above normal when hypertension is produced (fig. 1).

This resistance of the renal vessels to the vasoconstricting effect of angiotensin has been termed tachyphylaxis but this gives no indication of how it develops. It is associated with a fall in urinary osmolality on the infused side only (18) and with an increase in the excretion of kallikrein (20). The latter enzyme, by release of kinin within the kidney, may actively antagonise the action of angiotensin on the vasculature. Thus the renal clearances are maintained or may be increased by the action of the hypertension when the renal vessels are resistant to the vasoconstricting action of angiotensin. This may contribute to the natriuresis.

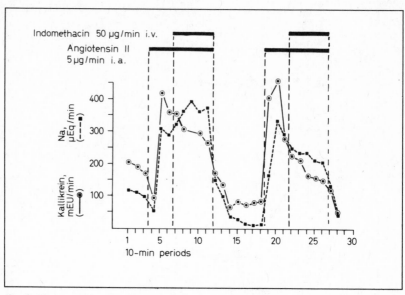

Fig. 2. The excretion of sodium (■ μEq/min) and kallikrein (⊙ milli-esterase units/min) from the left kidney prior to, during and after the infusion of angiotensin intra-arterially at 5 μg/min and superimposed indomethacin at 50 μg/min intravenously on two occasions in one dog.

When infusion of indomethacin at 50 μg/min i.v. was superimposed on the arterial infusion of angiotensin II at 5 μg/min there was a steady fall in the excretion of kallikrein (fig. 2). The sodium excretion either rose slightly or fell more slowly than the kallikrein when prostaglandin synthesis was inhibited by indomethacin. This indicates that the release of kallikrein which is produced by arterial infusion of angiotensin II depends upon prostaglandin as an intermediate. *McGriff et al.* (22) have shown that angiotensin stimulates release of prostaglandin of the E series and we may presume that it is this which stimulates kallikrein release.

The fact that sodium excretion does not follow the changes in kallikrein excretion but falls promptly when angiotensin infusion is stopped, suggests that the natriuretic mechanism initiated by the release of kallikrein, does not depend entirely on kallikrein for its continuation.

The super-imposition of noradrenaline in a dosage of 1 μg/min into the same renal artery caused almost immediate renal vasoconstriction with a fall in renal clearances (fig. 3). (31). This shows that the renal vessels which were resistant to angiotensin were still sensitive to the vasoconstricting action of noradrenaline. Since both aniotensin and noradrenaline cause release from the kidney of

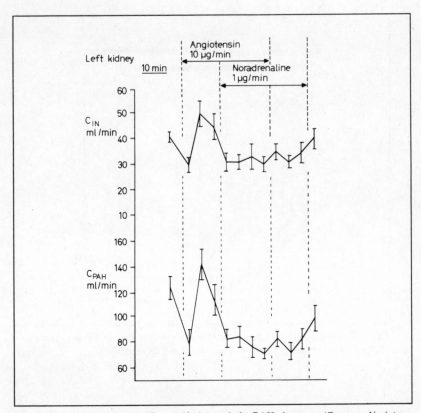

Fig. 3. The inulin clearance (C_{IN}, ml/min) and the PAH clearance (C_{PAH}, ml/min) on the left side prior to, during and after infusion of angiotensin intra-arterially at 10 μg/min and during the super-imposition of noradrenaline into the same renal artery at 1 μg/min. The initial point represents the mean and SE of two 10-min control periods in six experiments. Each of the following points is the mean and SE of six experiments and are sequential 10-min periods. From *Mills and Wilson* (31).

prostaglandin E (23), it is unlikely that this vasodilator is responsible for the tachyphylaxis to angiotensin.

With a fall in GFR as a result of noradrenaline infusion one might expect that sodium excretion would fall and it does (31). However, it does not fall as promptly as the GFR and so the timing of the effect of noradrenaline on angiotensin natriuresis is better shown by the sodium excretion as a percentage of that filtered. This is shown in figure 4. The fall off in natriuresis takes about 30–40 min after the noradrenaline infusion is added and so involves a mechanism other than change in GFR and which has a longer time course.

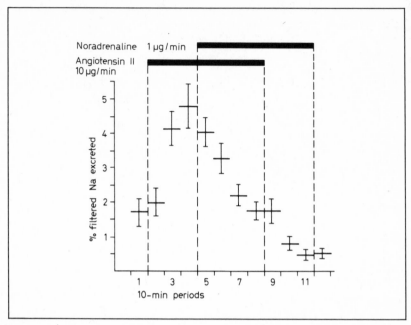

Fig. 4. The percentage of filtered sodium excreted for the same experiments shown in figure 3. Each point represents the mean and SE of the six experiments.

Noradrenaline and Kallikrein Release

Natriuresis has been correlated with the excretion of kallikrein in a variety of circumstances (27). Urinary kallikrein is of renal origin under normal circumstances (34). Acute arterial infusion of prostaglandin E_1 (36), acetylcholine (37), dopamine (28) and bradykinin (30) all produce natriureses correlated with kallikrein excretion. So also does rapid saline infusion in the dog (8) and release of renal artery constriction (4).

When noradrenaline was infused into the left renal artery in the dog at 10 ng/min there was a significant rise in sodium excretion from both the left and right kidneys (29). The same was true of urinary flow rate. When the noradrenaline was stopped the sodium excretion rose further and became highly significant: the same occurred with the urinary flow rate. The excretion of kallikrein during the infusion increased on both sides but only that from the right kidney was significant (table I). On stopping the noradrenaline infusion there was a much greater rise in kallikrein from both kidneys and this was very highly significant.

Table I. Effect of 10 ng/min noradrenaline infusion i.a. on urinary kallikrein excretion and renal function before and after prostaglandin synthesis inhibition (from *Obika*, 35)

		Before			After prostaglandin synthesis inhibition		
		C	E	R_c	C	E	R_c
Urinary	L	198.7	206.7	242.8***	151.4	150.3	150.2
kallikrein		± 6.2	± 7.3	± 8.8	± 2.6	± 6.1	± 4.1
excretion	R	171.9	195.8*	222.2***	150.0	127.3**	133.0**
mEU/min		± 5.6	± 6.0	± 5.6	± 2.0	± 5.7	± 4.1
UV, ml/min	L	0.70	0.82*	1.24***	0.51	0.48	0.45
		± 0.01	± 0.04	± 0.11	± 0.02	± 0.02	± 0.02
	R	0.53	0.64*	0.98***	0.42	0.42	0.46
		± 0.01	± 0.04	± 0.09	± 0.01	± 0.01	± 0.01
$U_{Na}V$, μEq/min	L	108.3	126.1*	148.1***	120.6	118.9	114.6
		± 2.8	± 5.5	± 3.1	± 1.5	± 2.5	± 7.8
	R	82.6	93.5*	120.4**	92.2	87.4	78.9
		± 1.8	± 4.6	± 1.7	± 13.2	± 1.2	± 2.5
U_KV, μEq/min	L	37.8	49.9***	57.3***	51.4	55.3	58.2
		± 1.4	± 3.3	± 2.3	± 2.9	± 2.4	± 1.8
	R	40.7	52.3***	57.8***	48.4	46.5	47.4
		± 1.7	± 1.8	± 2.4	± 3.8	± 1.3	± 1.4
Urinary	L	643	592*	495***	1,138	1,357***	1,363***
osmolality		± 5	± 22	± 31	± 41	± 33	± 4
mOsm/kg	R	688	655	466***	1,109	1,171**	1,293***
		± 9	± 21	± 46	± 16	± 8	± 32
C_{IN}, ml/min	L	31	35	41*	36	32	35
		± 1	± 2	± 3	± 2	± 2	± 2
	R	33	38	51***	33	27	30
		± 1	± 1	± 3	± 2	± 3	± 2
C_{PAH}, ml/min	L	54	60	77**	68	65	70
		± 4	± 5	± 7	± 3	± 3	± 3
	R	54	53	74**	63	54	61
		± 3	± 2	± 5	± 5	± 6	± 5

Values are means ± SE of six experiments in each case.
C = Control, E = experimental, R_c = recovery.
* p <0.05; ** p <0.01; *** p <0.001.

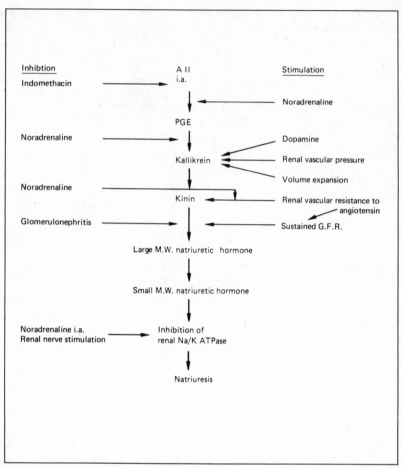

Fig. 5. The natriuretic chain of hormones originating from arterial infusion of angiotensin II is shown in the middle of the figure. On the left are inhibitory factors and on the right are stimulatory factors.

After blockade of α-receptors by infusion of phentolamine at 10 µg/min for 30 min, the infusion of 10 ng/min of noradrenaline had no effect on either kidney during the infusion or after it was stopped. Blockade of prostaglandin synthesis with indomethacin intraarterially at 2 µg/min also prevented the diuresis and natriuresis during and after noradrenaline infusion at 10 ng/min (table I). Similarly kallikrein excretion was not increased during or after noradrenaline infusion into the left renal artery after α-adrenoreceptor blockade or after inhibition of prostaglandin synthesis.

These data indicate that infusion of noradrenaline at 10 ng/min causes release of kallikrein and increased excretion of sodium and urinary volume. When prostaglandin synthesis was inhibited there was no increase in excretion of sodium, urinary volume or kallikrein. This indicates that the noradrenaline increases the secretion of prostaglandin, as *McGiff et al.* (23) showed previously, and that the release of prostaglandin was essential for the stimulation of kallikrein excretion.

Since the diuresis, natriuresis and excretion of kallikrein were all greater after the noradrenaline infusion was stopped and did not occur when prostaglandin synthesis was inhibited, it must be assumed that the noradrenaline itself was in part inhibiting the response which was due to the prostaglandin released by the noradrenaline.

It was indicated above that the resistance of the renal vessels to the vasoconstricting action of angiotensin was unlikely to be due to the prostaglandin released. However, since the data now indicate that noradrenaline inhibits the release of kallikrein and that release of kallikrein is dependent upon prostaglandin, the intra-renal relationship must be as shown in figure 5.

Other Activators of Kallikrein Release

The renal kallikrein/kinin system occupies a central role in natriuresis. Mechanical changes in pressure in the renal artery produce corresponding changes in the excretion of kallikrein (4). Sudden release of renal artery constriction causes a rise in kallikrein excretion. However, rises in blood pressure for other reasons will only increase kallikrein excretion if the mechanism of the hypertension does not cause a comparable increase in renal artery resistance. Infusing noradrenaline at 1 μg/min i.a. does not produce a constant change in kallikrein excretion though it usually produces some increase in blood pressure (35).

Similary vasodilators produce an increase in kallikrein excretion which may depend upon intra-renal pressure changes. When bradykinin was infused into the renal artery while a constriction reduced the pressure to 85–90 mm Hg, there was no natriuresis and no increase in kallikrein excretion (30).

Rapid volume expansion with saline in the dog caused a marked rise in kallikrein excretion which was correlated with the sodium excretion (27). However, wrapping one kidney in latex impaired the natriuresis and the kallikrein excretion without interfering with the normal rise in inulin and PAH clearances (21, 27). The direct pressure mechanism may, therefore, operate by changing a transmural pressure gradient.

Saline loading produces an increase in dopamine excretion which is related to the natriuresis (1, 11). The origin of the dopamine is not clear but it appears not to be filtered from the plasma. When dopamine was infused into the renal artery it produced a natriuresis which correlated with the increase in kallikrein excretion (28). The effect was not blocked by α- or β-adrenergic blockers but was inhibited by haloperidol which blocks the specific dopaminergic receptors.

The effect of volume expansion does not depend entirely upon prostaglandin because it is still effective when prostaglandin synthesis is inhibited (5, 39).

Kallikrein Stimulation without Natriuresis

Stimulation of kallikrein excretion is not associated with natriuresis under two different sets of circumstances. *Margolius et al.* (24) showed in man that administration of furosemide while subjects were on a low salt diet led to a marked rise in plasma renin and kallikrein excretion but with intense sodium retention after the furosemide was stopped. This suggested that the kallikrein/kinin system must be followed by another mechanism which in the salt-depleted state can be altered to inhibit natriuresis.

The second circumstance in which kallikrein can be stimulated without natriuresis was shown by *Godon and Damas* (15). In rats made to produce glomerulonephritis by immunisation with renal basement membrane, they showed that they had a normal increase in kallikrein excretion during volume loading but without natriuresis. The kidneys of animals which have been loaded with saline produce a substance which can facilitate natriuresis when the salt-loaded kidney is transplanted into the neck of a sodium deprived animal (33). *Godon* (13) has shown that renal tubular fragments from a salt-loaded animal produce this natriuretic substance in tissue culture, but it is not produced by renal tubular fragments of sodium-deprived animals. The kidneys of the glomerulonephritic rats did not produce this natriuretic substance. This substance has a molecular weight of about 40,000 and is clearly different from renal kallikrein.

A smaller molecular weight natriuretic substance has been identified in renal extracts of salt-loaded animals by *Gonick and Saldanha* (16) and by *Favre et al.* (12). This substance was not identified in kidneys from animals not salt loaded. One of its properties is inhibition of Na/K ATPase (17). Since it does not affect potassium excretion it must act on the more distal part of the nephron.

Stimulation of the renal nerves has been shown to decrease sodium excretion even without a fall in renal clearances (3, 40). It seems likely that the renal nerve stimulation would operate in a similar way to noradrenaline when it acts beyond the point of the kallikrein/kinin system. Since noradrenaline activates Na/K ATPase and can overcome the inhibition of this enzyme by ouabain (19), it seems possible that renal nerve stimulation, or noradrenaline itself, might decrease sodium excretion by overcoming the inhibition of Na/K ATPase which the substance described by *Hillyard et al.* (17) is known to produce.

The complete natriuretic hormone chain could be represented as in figure 5.

Conclusions

The natriuretic mechanism in the kidney is essentially a chain of events which can be inhibited or stimulated at a number of points. Starting with angiotensin II infused intra-arterially in a dose which is natriuretic, the first step is the release of prostaglandin E (PGE) which leads to release of kallikrein. Indomethacin can prevent the release of PGE and hence the release of kallikrein. Noradrenaline inhibits the stimulation of kallikrein release by PGE. Renal kallikrein releases lysyl-bradykinin or bradykinin which makes the renal vessels resistant to angiotensin and so maintains GFR. Noradrenaline overcomes this effect of PGE and leads to renal vasoconstriction.

Kallikrein release by the kidney can be stimulated, without PGE acting as an intermediate, by dopamine, by increased renal vascular pressure or by volume expansion. Kinins released in the renal cortex are probably responsible for the activation of the renal large molecular weight natriuretic hormone described by *Godon and Damas* (15). This release is prevented by glomerulonephritis in animals (15) and man (14).

The large molecular weight natriuretic hormone may be responsible for the release of the renal small molecular weight natriuretic hormone described by *Gonick and Saldanha* (16), *Hillyard et al.* (17), *Bricker et al.* (6) and by *Favre et al.* (12). It has been extracted from kidneys of salt-loaded rats and may be similar to the small molecular weight natriuretic substance described by *Clarkson et al.* (7) in the urine of salt-loaded human subjects.

Since *Hillyard et al.* (17) have described the effect of their small molecular weight substance inhibiting renal Na/K ATPase, it is likely that this is the last link in the chain of events leading to natriuresis. Na/K ATPase is stimulated by noradrenaline (19, 38) and this is likely to be one of the mechanisms by which renal nerve stimulation decreases sodium excretion (3, 40).

Summary

Large doses of angiotensin when infused intravenously or into the renal artery cause natriuresis. The initial effect is release of prostaglandin (probably PGE) and this leads to release of kallikrein. This latter step can be inhibited by noradrenaline. Activation of the kallikrein/kinin system is followed by release of a large molecular weight natriuretic hormone which is absent in glomerulonephritis. A small molecular weight hormone follows the large one and probably effects natriuresis by inhibition of renal Na/K ATPase. This inhibition is reversed by noradrenaline or renal nerve stimulation. Natriuresis is the result of a chain reaction and not a single specific natriuretic hormone.

References

1 Alexander, R.W.; Gill, J.R., jr.; Yambe, H.; Lovenberg, W., and Keiser, H.R.: Effects of dietary sodium and of acute saline infusion on the interrelationship between dopamine excretion and adrenergic activity in man. J. clin. Invest. 54: 194–200 (1974).

2 Barraclough, M.A.; Jones, N.F., and Marsden, C.D.: Effect of angiotensin on renal function in the rat. Am. J. Physiol. 212: 1153–1157 (1967).

3 Bello-Reuss, E.; Trevino, D.L., and Gottschalk, C.W.: Effect of renal sympathetic nerve stimulation on proximal water and sodium reabsorption. J. clin. Invest. 57: 1104–1107 (1976).

4 Bevan, D.R.; Macfarlane, N.A.A., and Mills, I.H.: The dependence of urinary kallikrein excretion on renal artery pressure. J. Physiol., Lond 241: 34–35P (1974).

5 Bohan, L.D. and Wesson, L.G.: Effect of prostaglandin inhibition on renal response to volume expansion in dog. Life Sci. 19: 1015–1022 (1976).

6 Bricker, N.S.; Schmidt, R.W.; Favre, H.; Fine, L., and Bourgoignie, J.J.: On the biology of sodium excretion: the search for a natriuretic hormone. Yale J. Biol. Med. 48: 293–303 (1975).

7 Clarkson, E.M.; Raw, S.M., and Wardener, H.E. de: Two natriuretic substances in extracts of urine from normal man when salt-depleted and salt-loaded. Kidney int. 10: 381–394 (1976).

8 Bono, E. de and Mills, I.H.: Simultaneous increases in kallikrein in renal lymph and urine during saline infusion. J. Physiol., Lond. 241: 127–128P (1974).

9 Wardener, H.E. de; Mills, I.H.; Clapham, W.F., and Hayter, C.J.: Studies on the efferent mechanism of the sodium diuresis which follows the administration of intravenous saline in the dog. Clin. Sci. 21: 249–258 (1961).

10 Fagard, R.H.; Cowley, A.W., jr.; Navar, L.G.; Langford, H.G., and Guyton, A.C.: Renal responses to slight elevations of renal arterial plasma angiotensin II concentration in dogs. Clin. exp. Pharmac. Physiol. 3: 531–538 (1976).

11 Faucheux, B.; Buu, N.T., and Küchel, O.: Effects of saline and albumin on plasma and urinary catecholamines in dogs. Am. J. Physiol. 232: F123–F127 (1977).

12 Favre, H.; Hwang, K.H.; Schmidt, R.W.; Bricker, N.S., and Bourgoignie, J.J.: An inhibitor of sodium transport in the urine of dogs with normal renal function. J. clin. Invest. 56: 1302–1311 (1975).

13 Godon, J.P.: Renal origin of the natriuretic material. Some chemical properties. Conference on Natriuretic Hormone, Bonn 1976 (in press).

14 Godon, J.P.: The oedematous phase of human glomerulonephritis is related to the disappearance of a natriuretic factor which reappears during recovery; in Moorhead, Baillod and Mion Dialysis, transplantation, nephrology, pp. 330–337 (Pitman Medical, Tunbridge Wells 1975).

15 Godon, J.P. and Damas, J.: The kallikrein-kinin system in normal and glomerulo-nephritic rats. Archs int. Physiol. Biochim. 82: 273–277 (1974).

16 Gonick, H.C. and Saldanha, L.F.: A natriuretic principle derived from kidney tissue of volume-expanded rats. J. clin. Invest. 56: 247–255 (1975).

17 Hillyard, S.D.; Lu, E., and Gonick, H.C.: Further characterisation of the natriuretic factor derived from kidney tissue of volume-expanded rats. Effects on short circuit current and sodium-potassium adenosine triphosphatase activity. Circulation Res. 38: 250–255 (1976).

18 Klein, G.L.; Mills, I.H., and Wilson, R.J.: Changes in renal function associated with the development of resistance of the renal vasculature to the arterial infusion of angiotensin. J. Physiol., Lond. 215: 43–44P (1971).

19 *Logan, J.G. and O'Donovan, D.J.:* The role of the (Na⁺-K⁺) ATPase in the uptake of
    noradrenaline. Ir. J. med. Sci. *145:* 307 (1976).
20 *Macfarlane, N.A.A.; Adetuyibi, A., and Mills, I.H.:* Changes in kallikrein excretion
    during arterial infusion of angiotensin. J. Endocr. *61:* lxxii (1974).
21 *Macfarlane, N.A.A.; Mills, I.H., and Bono, E. de:* Response to pressure changes of the
    renal kallikrein/bradykinin system in the excretion of sodium and water. Proc. Int. U.
    Physiol. Sci. *XI:* 366 (1974).
22 *McGiff, J.C.; Crowshaw, K.; Terrangno, N.A., and Lonigro, A.J.:* Release of a prosta-
    glandin-like substance into renal venous blood in response to angiotensin II. Circulation
    Res. *27:* suppl. I, pp. 121–130 (1970).
23 *McGiff, J.C.; Crowshaw, K.; Terrango, N.A., and Lonigro, A.J.:* Renal prostaglandins.
    Possible regulators of the renal actions of pressor hormones. Nature, Lond. *227:*
    1255–1257 (1970).
24 *Margolius, H.S.; Horwitz, D.; Geller, R.G.; Alexander, R.W.; Gill, J.R., jr.; Pisano, J.J.,
    and Keiser, H.R.:* Urinary kallikrein excretion in normal man. Relationships to sodium
    intake and sodium-retaining steroids. Circulation Res. *35:* 812–819 (1974).
25 *Mills, I.H.:* Regulation of sodium excretion: intra- and extra-renal mechanisms. J. R.
    Coll. Physns Lond. *4:* 335–350 (1970).
26 *Mills, I.H.; Wardener, H.E. de; Hayter, C.J., and Clapham, W.F.:* Studies on the afferent
    mechanism of the sodium chloride diuresis which follows intravenous saline in the dog.
    Clin. Sci. *21:* 259–264 (1961).
27 *Mills, I.H.; Macfarlane, N.A.A.; Ward, P.E., and Obika, L.F.O.:* The renal kallikrein-
    kinin system and the regulation of salt and water excretion. Fed. Proc. Fed. Am. Socs
    exp. Biol. *35:* 181–188 (1976).
28 *Mills, I.H. and Obika, L.F.O.:* The effect of adrenergic and dopamine-receptor
    blockade on the kallikrein and renal response to intraarterial infusion of dopamine in
    dogs. J. Physiol., Lond. *263:* 150–151P (1976).
29 *Mills, I.H. and Obika, L.F.O.:* A novel effect of intrarenal infusion of a non-vasocon-
    strictor dose of noradrenaline on renal function: relationship to renal kallikrein and
    prostaglandin. J. Physiol., Lond. *267:* 21–22P (1977).
30 *Mills, I.H. and Obika, L.F.O.:* Urinary kallikrein excretion during bradykinin and
    eledoisin infusions and its relationship to urinary osmolality. J. Physiol., Lond. *269:*
    72–73P (1977).
31 *Mills, I.H. and Wilson, R.J.:* Antagonism by noradrenaline of the changes in renal
    function associated with development of resistance of the renal vasculature to the
    arterial infusion of angiotensin. Proc. Int. U. Physiol. Sci. *XI:* Abstr. No. 1167
    (1971).
32 *Navar, L.G. and Langford, H.G.:* Effects of angiotensin on the renal circulation; in *Page
    and Bumpus* Handbook of experimental pharmacology, vol. 37, pp. 455–474
    (Springer, Berlin 1973).
33 *Nizet, A.:* Excretion of sodium and water by kidneys *in situ* and by transplanted
    kidneys following isotonic, hypotonic, iso-oncotic and hyperoncotic intravenous infu-
    sions in sodium-loaded and sodium-deprived dogs. Archs int. Physiol. Biochim. *84:*
    997–1015 (1976).
34 *Nustad, K.:* Relationship between kidney and urinary kininogenase. Br. J. Pharmacol.
    39: 73–86 (1970).
35 *Obika, L.F.O.:* The effect of some vasoactive agents on renal kallikrein in dogs and
    rats; thesis Cambridge (1977).
36 *Obika, L.F.O. and Mills, I.H.:* Effect of arterial infusion of prostaglandin E₁ on the
    release of renal kallikrein. J. Endocr. *69:* 45–46P (1975).

37 *Obika, L.F.O and Mills, I.H.:* The effect of acetylcholine on renal excretion of kallikrein, sodium and water (unpublished observations).
38 *Sulakhe, P.V.; Sheue-Heng, J., and Sulakhe, S.J.:* Studies on the stimulation of (Na^+-K^+) ATPase of neural tissues by catecholamines. Gen. Pharmac. *8:* 37−41 (1977).
39 *Susic, D. and Sparks, J.C.:* Effects of aspirin on renal sodium excretion, blood pressure, and plasma and extracellular fluid volume in salt-loaded rats. Prostaglandins *10:* 825−831 (1975).
40 *Zambraski, E.J.; DiBona, G.F., and Kaloyanides, G.J.:* Specificity of neural effect on renal tubular sodium reabsorption. Proc. Soc. exp. Biol. Med. *151:* 543−546 (1976).

Prof. *I.H. Mills,* Department of Medicine, University of Cambridge, Addenbrooke's Hospital, Hills Road, *Cambridge CB2 2QQ* (England)

Subject Index